THE POLITICS OF BAVARIA AN EXCEPTION TO THE RULE

For
Renate, Rich and Chris

The Politics of Bavaria – an Exception to the Rule

The special position of the Free State
of Bavaria in the New Germany

By

Dr Peter James

(University of Northumbria at Newcastle)

Avebury

Aldershot • Brookfield USA • Hong Kong • Singapore • Sydney

Published by
Avebury
Ashgate Publishing Limited
Gower House
Croft Road
Aldershot
Hants GU11 3HR
England

Ashgate Publishing Company
Old Post Road
Brookfield
Vermont 05036
USA

British Library Cataloguing in Publication Data

James, Peter
    Politics of Bavaria: Guide to the
    Political System of Germany's Largest
    Federal State
    I. Title
    320.9433

    ISBN 1 85972 166 4

Library of Congress Catalog Card Number: 95-78043

Design and typeset, in Palatino,
by Design Services, Department of Corporate Affairs
University of Northumbria at Newcastle

Printed and bound by Athenæum Press Ltd.,
Gateshead, Tyne & Wear.

Contents

List of Figures and Tables   vii

Abbreviations   ix

Foreword   xi

Acknowledgements   xiii

1. Introduction: Two Key Events in Recent Developments   1

2. The Historical Perspective   21

3. The Postwar Bavarian Political System   45

4. A new Constitution for a "new" State   67

5. Parties and Elections in Postwar Bavaria   87

6. The Development of the Bavarian Party System 1945-95   111

7. Bavaria's Booming Economy:
   From an Agricultural to a Modern Industrial State   129

8. A Study of Electoral Behaviour in Bavaria in 1990   147

9. Franz Josef Strauß – an Extraordinary Phenomenon   167

10. Bavarian Politics in the mid-Nineties   197

Appendices   209

References   213

List of Figures and Tables

Figures

3.1   The three traditional zones of Old Bavaria   47

3.2   The Development of the Bavarian Party System   66

5.1   The Franconion Protestant "corridor" in Catholic Bavaria   97

6.1   Bavarian vote on the Basic Law 1949   114

Tables

4.1   Denominational Breakdown within Bavaria   96

10.1 Results of the June 1994 European Elections in Bavaria   199

10.2 Results of the June 1994 European Elections in Germany   200

10.3 Results of the Bavarian State Elections in September 1994   202

10.4 Results of the 1994 Federal Elections in Bararia   204

10.5 Results of the 1994 Federal Elections in Germany   205

Abbreviations

AL	Alternative Liste
AUD	Aktionsgemeinschaft Unabhängiger Deutscher Action group of Independent Germans
B 90	Bündnis 90 Alliance 90
BfB	Bund freier Bürger Association of Free Citizens
BHE	Bund der Heimatlosen und Entrechteten Refugees' Party
BP	Bayernpartei Bavarian Party
CDU	Christlich Demokratische Union Christian Democratic Union
CSU	Christlich-Soziale Union Christian Social Union
DSU	Deutsche Soziale Union German Social Union
DVU	Deutsche Volksunion German People's Union
FDP	Freie Demokratische Partei Free Democratic Party
FRG	Federal Republic of Germany
GAL	Grüne Alternative Liste Green Alternative List
GDR	German Democratic Republic

JU	Junge Union Young Conservatives
KPD	Kommunistische Partei Deutschlands German Communist Party
NPD	Nationaldemokratische Partei Deutschlands National Democratic Party of Germany
NSDAP	Nationalsozialistische Deutsche Arbeiterpartei National Socialist German Workers' Party
PDS	Partei des Demokratischen Sozialismus Party of Democratic Socialism
SED	Sozialistische Einheitspartei Deutschlands Socialist Unity Party of Germany
SPD	Sozialdemokratische Partei Deutschlands Social Democratic Party of Germany
USPD	Unabhängige Sozialdemokratische Partei Ds Independent Social Democratic Party of G
WAV	Wirtschaftsaufbau Verein Economic Reconstruction League
Z	Zentrumspartei Centre Party

Foreword

There has never been a more urgent time to re-assess the position of the Free State of Bavaria than now. Bavaria was often, quite rightly, considered a special case amongst the West German Länder. It has its own special and enduring traditions, a unique political system and a very long and eventful history which has moulded Bavaria's sense of being a close-knit, self-contained unit with its own identity and idiosyncratic characteristics. This keenly developed feeling of existing as an independent nation state long before Germany existed created certain tensions in the political system of the Federal Republic between 1949 and 1990.

In the new Germany, against the background of the changing agenda of German politics, Bavaria faces different problems and new challenges. Two important events, which occurred within two years of each other to the day, meant that Bavaria was forced to reconsider its position fundamentally within the German polity. If Bavaria is not to lose the power and influence it acquired over many centuries, it needs to adapt quickly to a new situation in Germany, which is still evolving.

The purpose of writing this book is two-fold. Firstly, the many changes produced by German Unity have had a particular effect on the special position of Bavaria in the Germany's political system. The CSU, a unique phenomenon in the German party system, has been affected by developments in the new Germany and the party now faces new challenges in the post-Strauß era, after three decades of consolidation and stability in Bavarian politics.

Secondly, the book aims to present to the reader, to whom German sources are not available, an explanation of how and why Bavaria has so often behaved as a "rebel", taking an independent stance in German politics. Much of the material has not been presented in English before. Whilst a historical overview is provided, in order to assist the reader to understand the significance of Bavaria's long and enduring traditions, the book concentrates on the current situation, in which Bavaria and the Bavarians find themselves in the new Germany.

Acknowledgements

I wish to acknowledge the financial assistance of the German Academic Exchange Service (DAAD) which enabled me to make several vists to Bavaria in order to carry out my research.

I would also like to acknowledge the support of my former Head of Department, Dr Peter Hubsch.

Dr Peter James

1

Introduction:
Two Key Events in
Recent Developments

In the new Germany Bavaria today is "simply" one of sixteen
federal states in the FRG. That is the official position. However,
anyone who has examined the rich history and enduring political
traditions of Bavaria more closely knows that this is far from the full
story. The Free State of Bavaria constitutes a special case, worthy of
special study. However, during recent years new developments have
meant that there has never been a more urgent need to review the
position of Bavaria in the new Germany than now.

As has so often been the case in the past, Bavaria adopted a
Sonderstellung – a special position – in the political system of the Federal
Republic mark one, as Peter Pulzer calls it[1]. The modern federal state of
Bavaria has often displayed non-conformist qualities, behaving as a "rebel"
or "renegade", during the course of its long and eventful history. Bavaria,
precisely because it does constitute a special case, has frequently proved to
be the exception to the rule in the German political system.

---

[1] Prof Pulzer has referred to the "old" FRG as the Federal Republic mark one, and
to the new, united Germany as the Federal Republic mark two. He used these
terms again in a paper delivered to the Annual Conference of the Association for
the Study of German Politics at Corpus Christi College Oxford in April 1993.

This characteristic continued after 1949 and is still in evidence today in the Federal Republic mark two. In one sense the Free State of Bavaria has operated almost as a system within a system. It has, since 1949, clearly been an integral part of the FRG, whilst at the same time never missing an opportunity to underline its independent stance on certain issues.

Dominance of Strauß and the CSU

Just as Franz Josef Strauß was both a German (national) politician as well as a Bavarian (regional) one, so Bavaria's main political party – the Christian Social Union (CSU) – was, and still is, both a federal (national) and a Bavarian (regional) party. In the current situation, mainly as a result of two key events since 1988, there was, for a while, a real danger that Bavaria in general, and the CSU in particular, could have lost some of its former influence and power in the political system of the new Germany.

The achievements of the present Bavarian state are inextricably linked to the meteoric rise of a unique political party, the Christain Social Union, in post-war German and Bavarian politics. The particularly spectacular performance of the CSU during the period 1958-88 was, in turn, closely associated with the ebullient personality of Franz Josef Strauß, its most dynamic leader since the party was founded in 1945/46. However, two major events have occurred since 1988.

These events have, for the first time in over three decades, cast doubt both upon the omnipotent position of the CSU as both a national and a regional party in Germany, as well as on the position of Bavaria in the new Germany. The CSU is unique in the German party system, because it is the only party which has managed to adopt a political dual role[2]. This dual role, or *politische Doppelrolle*, whereby the CSU established two power bases, one in Munich and one in Bonn, will be discussed in more detail

---

[2] This role is referred to by Professor Alf Mintzel, Passau University, as "die politische Doppelrolle der CSU". Mintzel has published widely on the CSU. These publications are mentioned later when the CSU is examined in more detail. In his chapter on the CSU in Bavaria in Mintzel/Oberreuter. Parteien in der BRD. Leske und Budrich. Opladen. 1992, p.223, Mintzel says: "Die landes- und bundespolitische Stoßkraft und die Wirkung der CSU resultierten . . . aus ihrer institutionellen und politischen Doppelrolle als autonome Landespartei mit besonderem Bundescharakter."

2

later. First let us turn our attention to the two most fundamental developments which have affected Bavaria in recent years.

Bavaria in the post-Strauß era

These two events occurred within exacty two years to the day of each other. On 3.10.1993 a memorial stone, over four metres high, was unveiled by the CSU party chairman, Theo Waigel, in Reisbach, near Landau in Lower Bavaria. The two events commemorated are the day of German Unity – 3.10.1990 – and the death of Franz Josef Strauß – 3.10.1988. The initiator of the idea of the granite memorial stone, symbolising the division and unity of Germany, was Erwin Huber, the General Secretary of the CSU, whose home town is Reisbach.[3]

Strauß was without question Bavaria's most popular, successful and ebullient post-war politician. At the same time he was arguably Germany's most controversial one. Inside Bavaria he was the uncrowned "king", who could do almost no wrong; outside Bavaria he was regularly at the centre of political "affairs" and controversy, both at home and abroad. Strauß fired different emotions in different people. His opponents considered him unpredictable and unscrupulous. His supporters thought he was the epitomy of the dynamic and decisive politician.[4]

This stark contrast was never more in evidence than in the run-up to the 1980 federal election, when Strauß was the joint CDU/CSU candidate for the office of Federal Chancellor. This was the first and only time that the CSU has provided the Chancellor candidate so far. Despite his tremendous popularity in Bavaria, Strauß had to speak from behind a bullet-proof screen with police protection at some of his rallies in other parts of the country, for example in the Ruhr.

Large sections of the CDU and many of their supporters in North Germany found Strauß completely unacceptable as the Union parties' candidate. Many CDU voters north of the River Main thought that Helmut Schmidt was more suitable as German Chancellor, even though he was a

---

[3] The full deails are given in the Bayernkurier, Jahrgang 44/Nr 40. 9.10.93. p.1.

[4] See Parteien in Deutschland zwischen Kontinuität und Wandel. Koordination Emil Hubner, Heinrich Oberreuter. Bayerische Landeszentrale für politische Bildungsarbeit. München. 1992, p.63.

member of the SPD. The Schmidt/Strauß battle in 1980 was one of the worst examples of political mud-slinging in a federal election so far.

The strength of *anti*-Strauß feeling shown in many parts of Germany was matched only by the fanatical support *for* him in his home state, Bavaria. From 1961 onwards, when at the age of forty-five he became Party Chairman of the CSU, Strauß began to consolidate his position in the Bavarian political arena. He was determined that his party would never allow itself to be forced into opposition in Munich again, following the disaster, from the CSU's point of view, of the Coalition of Four (1954-57). This was the only Bavarian government from 1946 to the present day, which did not contain the CSU.

Strauß was equally determined to make sure that the Bavarian Party (BP), which in the early years of recovery after the second world war was a serious competitor for the CSU, could never recover from the resounding defeat the CSU had inflicted upon it. The BP was one of the four parties which opposed the CSU between 1954 and 1957. It was also the party which, from 1948 onwards embraced the Old Bavarian, Catholic/conservative farming and cultural traditions, so close the hearts of the rural community in Bavaria.

Even when Strauß was undergoing his most notorious and publicised defeat in German politics in the infamous Spiegel Affair in October 1962, his popularity in Munich was apparently unaffected. At the 1962 state elections in Bavaria, held on 25 November – only one month after the affair hit the headlines – the CSU increased its share of the vote, winning an absolute majority of the seats in the *Landtag* in Munich. Indeed, this was the start of CSU hegemony in the Bavarian party system, the achievement of which was, in no small measure due to Franz Josef Strauß.

On 30 November Strauß was forced to resign from his cabinet post as Defence Minister in Bonn, after admitting that he had lied to the *Bundestag*. This would have finished the career of many a politician. Not so with Strauß. In 1963 he was re-elected as CSU party chairman by over eighty per cent of his party's delegates, and he returned to power in Bonn as a minister only a few years later in the Grand Coalition government.

Record election results under Strauß

As minister president of Bavaria from 1978 until his death, ten years later, Strauß was virtually unassailable. He was simutaneously chairman of the CSU, a position he held from 1961 until 1988. It would hardly be an exaggeration to claim that from around 1970 to 1988 Strauß *was* Bavaria. He was certainly seen as a symbol, taken by many to represent and embody Bavarian interests. Bavarian elections began to bear a strong resemblence to presidential referenda on his personal popularity.

For example, in the 1974 Bavarian state elections the party which FJS had come to personify gained 62.1 per cent; in federal elections too the CSU recorded phenominal results on its home territory. In 1976 it polled 60 per cent. No other party in any region of (West) Germany has ever come close to attaining such phenomenal figures in either regional or federal elections. Such performances of course greatly strengthened the negotiating position of the CSU as a government coalition member.

At each Bavarian *Landtagswahl* in the Seventies and Eighties, the result was more or less a foregone conclusion. It seemed to be simply a statement of confirmation by the Bavarian people on the popularity of their party leader, who sometimes even predicted the percentage the CSU would gain before the election. The author witnessed this personally at the 1986 elections in Munich, when Strauß forecast 55 per cent "plus x" two days before the vote. On election day the CSU polled 55.8 per cent.

In fact the CSU election slogans in 1986 underlined their complete dominance of the political stage. With not a little arrogance, the party even appeared to be taking credit for the very high summer temperatures: *Sommer – Sonne – Bayern – CSU*. This was later followed by the simplistic phrase *Bei uns in Bayern*, with a picture of Strauß wearing a white shirt sitting in the sun (in 1982 they had used *Wir in Bayern*).

Another poster summed up the situation, as the CSU saw it, perhaps adequately for many of their voters: *Drei Namen, eine Kraft : Bayern, CSU und Strauß* (three names, one force), as if nothing more needed to be said. Although election posters and slogans should not be taken too seriously, these examples certainly spoke volumes about the CSU's perception of its own omnipotent position in Bavaria.

Strauß's sudden death from heart failure on 3.10.1988 was completely unexpected. Obviously such an abrupt departure from the political stage left a great vacuum. He had for so long – for nearly three decades, in fact – been so closely associated with Bavarian politics that the question of his successor had hardly even been considered, let alone seriously discussed. Certainly no "crown prince" had been chosen.

In his memoirs, which he began to record on tape in the autumn of 1987, Strauß himself mentioned briefly three possible successors as party chairman: Streibl, Waigel and Tandler. As a possible future minister president, he suggested either Streibl or Tandler.[5]

The two posts he had held for so long were in fact filled by Max Streibl, as Bavarian *Minister Präsident*, and Theo Waigel, who became *Parteivorsitzender*. Waigel, as well as being party chairman of the CSU also became Minister of Finance in Bonn. Gerold Tandler's reputation was tarnished after allegations of involvement in a financial scandal (the Zwick affair). Before that it certainly looked as if Strauß himself was grooming Tandler, always a close confidant, for high office.

At the time of the initial shock which many Bavarians clearly felt at the announcment of the death of FJS, two different forecasts of the future were generally expressed. Some felt that Bavaria without Strauß would never be the same again and they expected Bavaria's image to suffer accordingly; others thought that the modern, technocratic CSU party machine was so much in control that it could continue to fly on automatic pilot – an apt metaphor, perhaps, since Strauß was a keen amateur pilot.

The first test of public opinion came a few months later in June 1989. At the European elections the CSU share of the vote in Bavaria slumped from 57.2 per cent in 1984 to 45.4 per cent. A fall of that magnitude provided a real shock in itself, quite apart from the psychological shock of going from well above the magic fifty per cent mark to well below it. At the first elections to the European parliament in 1979, the CSU had polled 62.5 per cent.

Although the turn-out in 1989 was much higher than five years earlier, two worrying trends emerged for Bavaria's ruling party. Firstly, the loss of its absolute majority in Bavaria (viewed as catastrophic by CSU standards), and secondly a drop of almost twelve per cent from its 1984 result.

---

[5] Franz Josef Strauß. Die Erinnerungen. Siedler. Berlin. 1989, p.551.

Secondly, the right-wing Republican Party (*Republikaner*) gained 14.6 per cent of the Bavarian vote. Bavaria was the only federal state in the FRG where the REPS got into double figures. Their second highest result in the European elections in Germany was 8.7 per cent in Baden-Württembrerg, with an average German poll of 7.1 per cent. If a serious challenge to the hegemony of the CSU in Bavaria was to be mounted, it looked at that time more likely to come from the extreme right of the political spectrum than anywhere else.[6]

Strauß's oft-quoted remark that there must be no democratic, legitimate party to the right of the CSU[7] was recalled. Would Strauß have kept the Republicans on a tighter leash? It is not possible to do any more than speculate on such a question, although it certainly looked initially as if the populism of Strauß had, to some extent at least, been usurped by the demagogic Franz Schönhuber, the leader of the Republicans. It should in any case be remembered that the Republican Party, founded in 1983, dumbfounded all the opinion poll predictions by coming from nowhere to gain three per cent at the Bavarian state elections in 1986, when Strauß was still firmly in control.[8]

Nevertheless, such a significant drop in CSU support and a vote of nearly 15 per cent for a right-wing extremist party, facing a possible ban for being unconstitutional, only eight months after the death of Strauß was a serious setback. The total percentage vote for the SPD, Greens, FDP and REPS amounted to 50.6, compared with 45.4 for the CSU. Had the hegemony of the CSU been broken?

The xenophobic hysteria fuelled by some of the Republicans' statements clearly met with approval in certain quarters – and not just in Bavaria – at a time when the asylum question was a burning issue in German politics. Bavaria was a "front line" state in terms of receiving asylum seekers.

[6] "Wenn die CSU in ihrer Machtposition in Bayern gefährdet ist, dann von rechts, von den Republikanern." Interview with Professor Alf Mintzel at Passau University on 16.10.1993.

[7] op.cit Strauß, p.550. "Hierzu gehört meine oft widerholte Aussage und beschwörende Mahnung, daß es rechts von uns keine demokratische legitimierte Partei geben darf".

[8] This point was discussed by the author in an interview with Dr Franz Guber at the CSU headquarters in Munich on 15.10.1993. Dr Gruber felt it was impossible to say for certain that Strauß, had he lived, could have prevented the progress of the Republicans.

Alf Mintzel, who has written extensively on the CSU, states that there always have been right-wing groups which have not been fully absorbed[9]. The success of the far Right was dismissed at the time by the CSU as a protest vote; it was claimed that the electorate simply wanted to warn the Bavarian "party of government" against complacency and give them a warning, i.e. "something to think about" (*einen Denkzettel verpassen*). Yet the voters must have considered that there was something to protest about.[10]

Another major shock for the CSU

In the following year another surprise came. At the local elections (*Kommunalwahlen*) in March 1990 the CSU recorded its worst local election result since 1966. It not only polled over seven per cent less than in 1984 (in Bavaria local elections are held every six years), but the Bavarian SPD was remarkably successful in several town halls and local councils.

In Passau, an archetypal, ultra-conservative hotbed of Bavarian tradition, Willi Schmöller (SPD) took over as mayor from Hans Hösl, who had represented the CSU for 24 years. At the elections to the town council in Passau the CSU lost 7.3 per cent, dropping to 35.6 per cent, whilst the Republicans shot up from 5.2 per cent in 1984 to 11.3 per cent, becoming the third strongest party on the town council.

In Regensburg, another Catholic-conservative CSU stronghold, Christa Meier of the SPD became the first mayoress in Bavaria, replacing Friedrich Viehbacher. At the time of the initial shock, it was felt that if places like Passau and Regensburg, two of Bavaria's oldest and most traditional, die-hard bastions of conservatism, could have their town halls taken over by the SPD, then the CSU really was in trouble in the post-Strauß era.

Plattling, Straubing, Ansbach, Bad Kissingen, Füssen, Traunstein and Trostberg all gained mayors from the SPD, which took over control of 16 of the 25 county boroughs (*kreisfreie Städte*) in Bavaria. Before the local elections in 1990 the SPD had controlled only nine of them.

In the Bavarian capital the ruling mayor, Georg Kronawitter, who had

---

[9] Alf Mintzel. Political and Socio-economic Developments in the Postwar Era: the case of Bavaria, 1945-1989, p.175, in: Karl Rohe (ed.). Elections, Parties and Political Traditions. Berg. Oxford. 1990.

[10] Several articles in the Bavarian press at the time referred to "eine allgemeine Protest- und Denkzettelmentalität."

ousted Erich Kiesl, the only CSU mayor of Munich since the war, in 1984, increased his share of the vote by five per cent to 63.4. His challenger from the CSU, government spokesman Hans Klein, received 24.6 per cent. Ingrid Schönhuber, the wife of the Republican leader, gained 5.2 per cent.

Kronawitter stepped down three years later, which brought about a premature election. On 12.9.1993 Christian Ude (SPD) defeated the controversial CSU politician Peter Gauweiler. The result in Munich, always an SPD stronghold, was closer than usual – 52.1 per cent for the SPD candidate against 41.8 for the CSU (1990: 61.6 and 26.3 per cent respectively). Ude was a relatively unknown newcomer, and the defeat was still viewed by many as ignominious for the CSU, even though the party has only ever provided one mayor of Munich since the war (Kiesl in 1978).

At the 1990 local elections in Bavaria's second city, Nuremberg, the ruling SPD mayor, Peter Schönlein, also retained his position (56.3 per cent). The Bavarian minister president, Max Streibl, spoke of "an extremely difficult situation" for the CSU in the cities[11]. The down-trodden CSU found some consolation in the fact that the Bavarian SPD, despite its electoral success in Bavaria's town halls, had overall dropped just over two per cent of its vote in the local elections, which brought it down to below the thirty per cent barrier, which is often taken as a benchmark for the SPD in Bavaria.

The SPD successes were all the more surprising in the context of the party's weak performance in post-war Bavarian politics. The SPD's results have been consistently worse in Bavaria than in any other *Land*.

The reasons for such a surprisingly poor CSU performance in March 1990 were explained to the author[12] in the following terms. Whilst the death of Strauß was a factor, it was not considered *the* major factor. Local elections are, above all, a vote on personalities and their performance in the local community. Just one example of several will serve to illustrate the point that in many cases CSU mayors and local politicians had taken their voters for granted for too long. Complacency had crept in.

---

[11] "Eine außerordentlich schwierige Situation," reported in the newspaper Darmstadter Echo on 19.3.1990. p.1

[12] In interviews at the party headquarters with Dr Guber, in the Bavarian State Chancellory with Dr Baer and in discussions with Dr Hopfenmüller, director of the Hanns-Seidel-Stifung, all in Munich in October 1993.

In the Swabian town of Kempten, in the Allgäu, Josef Höß had been the CSU mayor for twenty years. He decided to build a controversial waste disposal unit close to the local community. The local population protested strongly. Höß told the people he was going ahead with construction plans regardless. In fact he eventually backed down just before the election, but by then it was too late. He had lost the people's confidence in him and suffered the consequences.

The way this incident and many others were handled was explained to the author as an example of how no party, not even the omnipotent CSU in Bavaria, can afford to take its voters for granted. It was felt in the CSU hierarchy that in many instances its losses could be attributed more to its own mistakes than to SPD achievements. Complacency and arrogance on the part of numerous CSU local officials and representatives were blamed for the party's very disappointing results at the 1990 local elections.

Another issue surfaced at the time of these elections. On the same day, 18 March 1990, elections were held for the East German *Volkskammer*. The CSU's partner organisation, the DSU, was having problems establishing itself and some sections of the CSU, including Gauweiler and Wolf, the Munich leader of the youth organisation JU (*Junge Union*), were calling for closer cooperation with the DSU. Strauß, it was claimed, would not have been as hesitant as the new CSU leadership in securing the party's future in the forthcoming united Germany. He would have moved into the East more quickly, as the other main parties had done.[13]

Whilst it is true that the other main parties were quicker off the mark to establish their counterparts in the East – this paid dividends for the FDP in the all-German elections in December 1990 – such a policy also had clear implications for the CSU's dual political role and its special position as a Bavarian party.

The point was made in May 1990 – the electoral system is complicated for local elections in Bavaria, and the full and final details of all the results were not announced until May – by Werner Kaltefleiter that if the CSU were to try to extend its influence to Saxony, Thuringia and other areas, it would be in danger of losing the dominant position it held. That position

[13] Speiegel no. 23/1990, p.27. A detailed report on the in intial trends which emerged in the Bvarian local elections is given in Spiegel no. 15/1990, p.135f.

of power and influence was based on its identification with Bavaria and vice versa.[14]

Bavaria adapts to the new Germany

With the approach of German Unity in early October (the Bavarian state elections were scheduled for 14.10.1990), the mood of the German people in East and West became euphoric. Important issues in German politics at the time (e.g. unemployment, asylum, the environment, housing), which under normal circumstances would have been the main campaign themes of the approaching federal election – the first all-German elections for 58 years – on 2 December, were subsumed into the hysteria of the one central theme which overshadowed everything else: re-unification.

Any problems the CSU might have been facing in a new post-Strauß era were forgotten, as everyone concentrated on the record number of other local, regional and GDR polls. When German Unity arrived, the federal CSU, as a member of the coalition government, derived some cudos from the euphoria and hysteria of the whole event.

Although it was the CDU and Kohl personally who took most of the credit (and later the blame, especially in the eyes of the East Germans), at the time the CDU/CSU/FDP coalition government was identified in positive terms with the historic achievement of German Unity on 3 October 1990.

Helmut Kohl was happy to bask in the glory of being the Federal Chancellor who had given the German people Unity (*Kanzler der Einheit*) and must have been delighted he no longer had to do constant battle with Franz Josef Strauß, who had made his political life so difficult for so long. The on-going battles between Kohl and Strauß, which always came to a head around the times of federal elections, left their mark on the CDU/CSU relationship. This important aspect of German coalition poli-

---

[14] Rolf Linkenheil reported the speech of Kaltefleiter, the Director of the Institut für Politische Wissenschaft of the University of Kiel in the Hannoversche Allgemeine newspaper on 18.05.90. Some of Kaltefleiter's remarks were reproduced in Aus Politik und Zeitgeschichte on 24.01.92, p.24 in an article by Hans-JügenLeersch: Die CSU: eine neue Form der Bayernpartei? "Die CSU gewinnt ihre Stärke durch ihre dominierende Rolle in Bayern, die auf der Identifikation dieser Partei mit Bayern und umgekehrt beruht. Eine Ausdehnung auf andere Länder . . . würde dieses bayerische Profil der Partei nur verwässern."

tics, in which Strauß and the CSU played a crucial role, is discussed in a later chapter.

The electric atmosphere in Germany in the autumn of 1990 affected the Bavarian elections, held only eleven days after the Unity celebrations. The CSU polled 54.9 per cent. Although this was in fact less than at any of the previous five state elections in Bavaria, it *was* the absolute majority which the new Bavarian leadership under Streibl and Waigel so desperately wanted.

For quite some time there really had been considerable doubt about whether the new CSU leadership in the post-Strauß era would be able to retain the absolute majority which everyone had simply taken for granted during the previous twenty years. In fact the 1990 CSU result was less than one per cent below the 1986 Bavarian state election result under Strauß. It was consequently greeted in Bavaria as an excellent performance, given that it had come only two years after the death of FJS.[15]

Many people breathed a huge sigh of relief when the Republicans' result of 4.9 per cent was announced. A poll of under five per cent at a Bavarian state election means no representation in parliament in Munich. In one sense, the ground had been taken from under the Republicans' feet, once Germany had been united, since Schönhuber and his supporters had often demanded the return of German territory in the East.

There was great disappointment in the headquarters of the Bavarian SPD, who were hoping to clear what for them had become a psychological thirty per cent hurdle. In the months before the first state election since the death of Strauß this seemed a realistic possibility. In the event, however, their result of exactly 26 per cent was even lower than their poor performance four years earlier (1986: 27.5 per cent). The Greens (6.4 per cent) gained twelve seats and the FDP (5.2 per cent) seven seats in the new Bavarian parliament. Four parties entered parliament. Nevertheless, with 127 out of 204 seats, CSU hegemony in Bavarian politics was confirmed again.

---

[15] On 15.10.1990 the Munich newspaper AZ reported: "Waigel und Streibl sind Strauß' Schatten los." On the same page, p.6, the report said that Germany was larger and Bavaria more Bavarian, but not "schönhuberisch." (The Republicans failed to clear the five per cent hurdle.)

Loss of influence for Bavaria in the new Germany

After such a convincing result it was no longer the hegemony of the CSU party machine in Bavaria that was being called into question. It was more a matter of Bavaria's position in the new Germany. The danger which has been identified for the CSU since the 1989/90 period is that it could become simply a Bavarian party, retaining its dominant role in Munich but losing its influential role as a federal German party.

On 3.10.1990 Bavaria went from being one of ten federal states in West Germany to one of sixteen in the new Germany, which enlarged its territory by almost thirty per cent. Obviously that change had many implications, including of course primarily financial ones, as vast amounts of money were, and still are being pumped into boosting the dire economic situation in the new *Bundesländer*.

Several authors commented on the weakening of the CSU's position with re-unification. Eckhard Jesse observed that, although it was the most clearly anti-Communist of the established parties, the CSU room for manoeuvre outside its home territory of Bavaria was very limited.[16]

In the days of the Federal Republic mark one, Munich was often referred to as the "secret capital." Its magnificent architecture and impressive monuments, stemming from the various "golden ages" of the Wittelsbach rulers, give Munich the aura of a capital city. For many years Munich took on the mantel of being the capital of South Germany, with Hamburg as the capital of North Germany. Now Germany has a "new"capital – Berlin – and there is no longer any need to look for cities which have a greater claim to capital status than Bonn, whose temporary status had become somewhat permanent.

There is no doubt that the full financial implications of the cost of German Unity were not fully grasped until well after the event. Certainly they were not understood at the time of re-unification. Neither the main players on the political stage, nor the people in either the West or the East realised just what enormous political, economic, cultural and social problems were involved.

---

[16] "Die CSU ist durch den Prozeß der Wiedervereinigung geschwächt worden." Eckhard Jesse in Parteien in Deutschland zwischen Kontinuität und Wandel. Koorination Emil Hübner, Heinrich Oberreuter. Bayerische Landeszentrale für politische Bildungsarbeit. München. 1992, p.84.

Eva Kolinsky referred to some of these social problems which accompanied re-unification. Bavaria was a common crossing point for east Germans coming in to the West. The euphoria soon died down in border towns like Hof, as the Trabi cars continued to come and pollute the environment. It got so bad that erstwhile border villagers blockaded the road leading west from the former GDR.[17]

Although it remains the largest German federal state in terms of area (70,554 sq.km.), Bavaria now covers less than one fifth (19.7 per cent) of the territory of the Federal Republic. It used to cover well over one quarter (28.3 per cent). Bavaria formerly had five of the 45 votes in the *Bundesrat*, the Upper House in Bonn. In the new system it has less influence with six votes out of 68.

The denominational dimension

Cabinet minister Volker Rühe said that after re-unification Germany was larger, more northern, more eastern and more Protestant. Each of these factors could be considered a disadvantage for Bavaria.

Every examination of voting patterns in Germany, right down to local level, reveals that religious affiliation, and in particular **membership** of the Catholic or Protestant Church, has often been an important influence on how German citizens vote. The key factor is not simply denomination (*Konfession*), i.e. church membership but regular church **attendance**. This has always been one of the two main cleavages in determining electoral behaviour.[18]

Regular church-attenders in Germany, especially Roman Catholics, have always tended to support the Union parties. Protestants, who have a much

---

[17] Eva Kolinsky. Inaugural Lecture. People and Politics in the Unified Germanies; A Citizens' Germany in Europe? University of Keele. 15 October 1992, p.3. Kolinsky also states that unification has exploded the myth that but for the Berlin Wall the German people East and West had remained the same. This created social tensions in Bavaria too.

[18] The model of five electoral blocs, including the two main cleavages referred to, discussed in Stephen Padgett and Tony Burkett. Political Parties and Elections in West Germany. Hurst. 1986. p.285 was an excellent guide to German voting patterns in the West and still has a certain validity, in general terms today. The current situation has of course been subject to a number of developments and is extremely volatile, especially with regard to voters from the East.

14

smaller proportion of regular church-goers, especially if they are also trade union members show a clear preference for the SPD. Although the denominational dimension is nowadays a declining factor, it is still relevant.[19]

From around the mid-Fifties, a secularisation process began in West German society. This process was accelerated during the Sixties. Catholics were always much more regular church attenders than Protestants. In 1950 around 55 per cent of Roman Catholics attended mass regularly. That represented a large reservoir of likely CDU (and in Bavaria CSU) voters. This figure dropped to around 33 per cent after secularisation. For Protestants over the same period the figure has dropped from around 20 per cent regular attendance to around 10 per cent. The most recent figures available suggested that in 1993 only around one-fifth of Catholics and less than five per cent of Protestants in the new Germany go to church regularly.[20]

The vast majority of the population of the former GDR was, officially at least, Protestant – only seven per cent were members of the Roman Catholic Church. Under the SED regime, however, church membership and attendance were discouraged. It is therefore, strictly speaking, incorrect to state that after forty years of living in East Germany the majority of the population is Protestant, despite the role of the Churches (e.g. the Monday demonstrations in Leipzig etc.) in protest marches before the Wall fell.

For this reason the new statistics on the number of Protestants and Catholics in Germany should be treated with caution. In the former West Germany there was, from 1949 onwards, always an approximate balance between the two major denominations. In 1959 there were 25,476,000 members of the Catholic Church in the FRG, of whom 11,876,000 attended mass regularly. Of the 26,650,000 Protestants in 1959 around 1,300,000 were regular church-goers. In 1980 there were 26.7 million Catholics and 26.1 million Protestants.[21]

---

[19] "The cleavage line between the secular and the religious is thus still a major though declining factor in predetermining electoral choice in the Federal Republic." William E. Paterson and David Southern. Governing Germany. Blackwell. Oxford. 1991, p.180.

[20] In a report in Spiegel no 10 from 6 March 1995 "Der Himmel muß warten" (p.76) the figures of 19.3 and 4.8 per cent respectively for German Catholics and Protestants attending church regularly in 1993 were given.

[21] Figures taken from David Childs. Germany in the Twentieth Century. Batsford. London. 1991. p.304.

This approximate balance was confirmed in the 1987 census: 26,232,004 R.C. (42.9 per cent) and 25,412,572 Prot. (41.6 per cent). It must be remembered that these percentages are of the total population of the FRG in 1987 (61 million approx). In 1995 the new Germany had a total population of over 80 million, which included more foreigners – over 5.6 million, amongst whom there were 1.7 million Moslems (mostly Turks) – plus a trend amongst the German population itself towards secularisation.

In 1991, after German Unity, there were 28,198,000 members of the R.C. Church and 29,208,000 members of the Protestant Church[22]. In Germany these figures must be registered, because church taxes are deducted at source from the incomes of all Catholic or Protestant church members.

These figures indicate that, of the total number of members of the German Churches today (58 million approx.), some 48 per cent are Catholic and about 50 per cent are Protestant. If the numbers of registered church members is calculated as a percentage of the total popuation of Germany (approximately 79 million in 1991), then obviously lower figures emerge: 35 per cent R.C., 38 per cent Prot. These are probably more meaningful figures in the present secular climate, although the figures should be treated with great caution in any case.

Whilst there has been a slight increase in the number of Protestants in the new Germany, some more cynical voices say that Germany, rather than becoming more Protestant, has become more heathen, since it took in the five new states.[23]

In Bavaria, the precise statistics themselves are almost irrelevent, owing to the extremely strong and close ties between the Catholic Church and the Bavarian State, which were established many centuries ago. The 1987 census revealed a slight fall in the percentage of Roman Catholics in Bavaria (67.2 per cent, with 23.9 per cent Protestant) – compared to almost exactly 70 per cent Catholics from 1946 onwards. This minor change is, however, no more than could be expected in the current secular climate. In the Weimar Republic 73 per cent of the Bavarian population were Catholic.

---

[22] Figures taken from das Statistische Jahrbuch für die BRD 1993. Metzler Poeschel. Wiesbden. 1993.

[23] This point was put to the author in an interview with Wilfried Scharnagl, editor of the Bayernkurier, in Munich on 19 October 1993.

Bavaria's whole church history and enduring religious traditions reveal such close ties to the Church of Rome[24] that the links between regular attendance of the Catholic Church and voting CSU were virtually set in tablets of stone right from the start of the post-war Bavarian state.

Indeed, as a consequence of the very thorough organisation of the party technocrats in the modern CSU party machine, the party even succeeded in attracting substantial numbers of Protestant voters – especially the regular church attenders – when the CSU was in opposition in Bonn from 1969 onwards. This portion of the Bavarian electorate was traditionally a source of support for the SPD, and to a much lesser extent, the FDP.

In summarising the effects of these two key events on the beginning of a new era in post-war Bavarian politics, the author gained the clear impression, from interviews and research in Bavaria, that the establishment of the new Germany was of greater significance.

Whilst the loss of Bavaria's most effective and popular leader since 1945 obviously had a major influence on both German and Bavarian politics, the CSU limousine was so well serviced that, after the initial shock of the team's top driver (FJS) suffering a heart attack whilst at the wheel, it was able to continue running. Two new drivers (Waigel, Streibl) replaced the original one. One of these has already been replaced, (Stoiber replaced Streibl) but the party limousine is now running smoothly again.

Franz Josef Strauß was undoubtedly the driving force behind much of the CSU's great success. He was a past master at expanding and at the same time balancing the party's dual power base in Bonn and Munich. Nonetheless no-one is indispensable and no party can afford to invest its whole future in one leader. The CSU, wisely, did not do that. Thus it has managed to recover from the initial shock and electoral setbacks it suffered during the period between autumn 1988 and spring/summer 1990.

During his last few years Strauß, who was 73 when he died, was not beyond criticism in his own party[25]. To name just one example, his decision in 1988 to exempt aircraft fuel for private use (Strauß was himself a keen

---

[24] The historical perspective given in the following chapter explains the role played by clerics, monasteries and church life in Bavarian society.

[25] op. cit. Jesse in Parteien in Deutschland. p.63. "Jedoch nahm die Parteibasis besonders inden letzten Jahren vor seinem (Strauß) Tod die oft eigenmächitg

amateur pilot) from a general tax increase on aircraft fuel was controversial and difficult to understand. There were other criticisms too of a leader who, in the final stage of his life, was obviously not as physically and mentally fit as he had been. Indeed the author was told by one source in Bavaria that, in purely political terms, Strauß's death occurred at just the right time.

So, despite the loss of Bavaria's most popular and successful leader, the CSU was more than capable of carrying on. The structures and procedures were so firmly in place, partly as a result of Strauß's own efforts over the years, that the Bavarian party of state, marched forward.

It must be conceded, however, that Strauß's successor, Max Streibl, had to be replaced in June 1993. After the Amigo Affair first broke, the criticism of Streibl, which had been growing for some time, reached a head. It was claimed that Streibl was too much of a backwoodsman, hardly known outside Bavaria. There was little confidence that he was the right person to lead the Free State of Bavaria into the super election year 1994, when a total of twenty elections were due in Germany.

A younger, more dynamic man than Streibl (then aged 61) was required. Edmund Stoiber, ten years his junior, had already been tipped as a "high flyer", but counted as one of the younger generation when Strauß was alive. Stoiber was in charge of the State Chancellory from 1982-88; he had always been regarded as extremely ambitiousand and considered to be on the right wing of the party.

The general view initially, confirmed since, was certainly that Stoiber presented a more dynamic image for the CSU when he took over in June 1993. In typical CSU fashion Stoiber criticised the CDU on 30.12.1993 on two counts, as if to remind us, as FJS did periodically, that the CSU is an independent party and does not automatically agree with everything that its conservative coalition partner does. Specifically, Stoiber accused the CDU of not doing enough to integrate right-wing protest voters into the party, as well as being indistinguishable nowadays from the SPD.

Also in December 1993 the special relationship between the Free State of Bavaria and the Church of Rome was again apparent. Christmas gifts of Bavarian sausage (Weißwürste) and beer from the former monastery at Andechs were sent to the Vatican. Bavarian honey was also one of the gifts to the Pope and Cardinal Ratzinger, Bavaria's envoy in Rome, which were intended to underline the good relations which exist between the Bavarian state and the Vatican.

The changes brought about in October 1990 were fundamental and far-reaching. They have had a greater effect than the loss of Strauß per se on both the special position of Bavaria in the new German polity and on the position of the CSU. It is in this context that the most recent developments in Bavarian politics should be viewed. Nevertheless, the present situation in Bavaria, one of Europe's oldest states, can not be fully understood without being viewed in its historical context.

2

The Historical Perspective

Bavaria is one of the oldest states in Europe. Anyone who is unfamiliar with Bavaria's long and varied history will find it difficult to understand all the nuances of the modern state. It is essential, for example, to realise that Bavaria today is much more than simply one of sixteen federal states in the new Germany. Bavaria and the Bavarians often feel that they occupy a special position, owing to their special history and long and enduring traditions which developed over many centuries. This is recognised by Germans both inside and outside Bavaria with the phrase "things are different in Bavaria" (*in Bayern gehen die Uhren anders*). The expression conveys the idea that Bavaria is often the exception to the rule, not always wishing to conform. The Free State of Bavaria has, over the centuries, periodicaly evinced "rebel" characteristics, sometimes behaving as a "renegade".

The observer of present-day Bavaria will also hear frequent reference to the Bavarian independent, free spirit (*die bayerische Eigenständigkeit*). Before this can be fully understood, it is important to grasp the fact that Bavaria developed as a nation state several centuries before Germany officially existed as a unified country. In fact Bavaria evolved as a dukedom, a territorial unit and a kingdom long before German unification in 1871.

Many parts of present-day Bavaria can look back on a millenium of history. Indeed some towns and cities are two thousand years old. There were Germanic tribes living in the area of what is now Bavaria, Swabia and the Tyrol shortly before the birth of Christ. Archeological evidence confirms that Celts occupied the region which forms northern Bavaria today around that date.

As is so often the case, it helps to understand the present situation better if you understand the past. This is particularly true of Bavaria, which possesses an extremely long and rich history. A historical overview, however brief, is therefore indispensable, since it will then be much easier to grasp not just territorial changes but also the influences on life and society which have moulded attitudes and behaviour in modern Bavarian cultural and political life. The first thing to consider is who the Bavarians are and where they came from.

The Bavarian people

The eminent Bavarian historian, Karl Bosl, confirms that there were tribes living in areas of what is now Bavaria, Swabia and the Tyrol shortly before the birth of Christ. Settlements developed, particularly around the alpine foreland region and central Bavaria at points on rivers, for example the Danube, Vils, Pegnitz, Main. However very little is known about these early tribes. From around the 4th century AD it is known, however, that Celts occupied the territory and created the first towns, called "oppida", in places such as Passau, Regensburg, Kempten and Straubing. The Celts were the dominant influence until the Romans arrived. Any present-day Bavarian place names ending in -ing, for example Erding, Altötting, Roding etc., are amongst the oldest towns in Bavaria.

The Romans wanted to protect their lands in northern Italy, so they forced their way forward until they reached the river Danube. In order to secure their territory, the Roman armies established the provinces of Rätien and Noricum – the region which later became Bavaria – and built a Roman wall, the Limes, between the rivers Main and Danube. They set up camps in places which are today some of Bavaria's oldest towns. Castra Regina on the river Regen (Regensburg), Castra Batava (Passau) and the capital of Rätien, Augusta Vindelicum (Augsburg) were just three of the strategic points at which a Roman camp or "castell" was established.

The inhabitants of this alpine foreland region, which later developed into part of modern Bavaria, were a mixture of Celts, Romans and various Germanic tribes, known collectively as Romanic Celts. As in Britain and elsewhere, the Roman armies contained soldiers from Gaul and other areas who inter-married with the local population. Bosl states that archeological excavations reveal that even before the Romans arrived in this region a feudal system existed with a type of nobility, an agricultural elite and druids (priests).

The Roman Emperor Constantine was prepared to tolerate Christianity and from the end of the 4th century, under Emperor Theodosius, it was officially "permitted". From the middle of the 6th century onwards the inhabitants of this region were known as the *Baiovarii* or *Boiovarii* (modern terminology: *Bajuwaren*). They were also known by the shorter name of Boier, referring to Celts, later adapted to Baier. This was the origin of the modern German word for a Bavarian (modern spelling *Bayer*), and the name of the region, originally spellt *Baiern*, now *Bayern* (same pronunciation) is derived from this. The *Baiovarii* became the dominant tribe, first mentioned in the year 551 AD in Germanic tribal history, and from that point on the Bavarians, in modern parlance, exerted a considerable influence on this territory. They also gave it its name.

Bavaria as a dukedom

The history of Bavaria really begins with establishnent of a dukedom (*Stammesherzogtum*) in the second half of the 6th century under the rule of the Agilolfinger. The first ruler was Duke Garibald I (554-594). He ruled over the people referred to above, known as *Baiovarii* or *Bajuwaren*, whose language was probably Romanic and whose religion was officially at least Christianity.[2]

Tassilo I (595-610) was the next Duke, followed by Garibald II, who ruled until the year 640. During the period 738-42 Duke Odilo and Bonifatius (St. Boniface), a cleric, began to give some shape to religious life in Bavaria by organising the bishoprics of Regensburg, Passau, Freising, Salzburg, Würzburg and Eichstätt[3]. In some respects this may be regarded

---

[2] ibid, p.20.

[3] Information über Bayern, a brochure published by the Bayerische Staatskanzlei. Munich. 1991.

as the first attempt to give the area a separate, corporate identity by elevating it to a church province in its own right.

This rebel/renogade theme of wanting to establish a separate identity recurs at regular intervals throughout Bavaria's long and eventful history. At that time the Dukedom of Bavaria included Salzburg and Vienna in the East Mark (*Ostmark*). Bavaria in those days stretched south beyond Brenner and included Bozen/Bolzano, nowadays part of northern Italy. When Bavaria lost this territory in 976, the territory of the East Mark joined the area which later became Austria.

In 766 Duke Tassilo III founded a Benedictine monastery and a nunnery on the island of Chiemsee. In 788 Charlemagne was crowned in Rome by Pope Leo III and he became ruler of the Holy Roman Empire, known in German as *das Heilige Römische Reich deutscher Nation*. He incorporated *Baiern* into his empire.

Between 554 and 1623 Bavaria had over one hundred rulers, followed by 7 princes or electors (*Kurfürsten*) between 1623 and 1806. Numerous developments and changes occurred during such a long time-span, and obviously the many different rulers left their mark, to a greater of lesser extent, on the nation state which emerged. Even when the modern state of Bavaria was founded at the beginning of the 19th century, it was still over sixty years before the German nation state was founded.

Between the sixth and eigth centuries a Law of the Bavarians (*Lex Baiuvariorum*) was gradually developed. This is the oldest known Bavarian law book.

The growth of towns and cities

Around 1050 Nuremberg was founded, although at that time it was part of the separate dukedom of Franconia (*Franken*). Swabia (*Schwaben*) too was in those days an independent dukedom. In 1158 Henry the Lion founded Munich. He re-directed the lucrative salt route away from Freising by building a bridge over the river Isar. The first stone bridge in Europe was built across the river Regen. It was started in 1135, completed in 1146 and can still be seen today in Regensburg, one of Bavaria's oldest towns. Regensburg fulfilled the function of a capital until the middle of the 13th century.

In Franconia imperial cities (*Reichsstädte*), bishoprics (*Bistümer*) in Würzburg and Bamberg and markgravia (*Markgrafschaften*) in Ansbach and Bayreuth were created. In Swabia imperial cities, the archbishopric of Augsburg and the imperial abbey in Kempten were created[4]. During this time Bavarian territory was becoming rather fragmented.

Christianity, although present under Roman rule, really first came to Bavaria in the eighth century, brought by Irish and Scottish monks. They came to the area of Old Bavaria (*Altbayern*). It was here that the Roman Catholic faith was first established and developed over subsequent centuries. From 1070 the Welfen ruled as Dukes of Bavaria, and at approximately the same time the Staufer were raised to the status of Dukes in Swabia, which resulted in some competition. The close links between Bavaria and the Catholic Church were further strengthened when three Bavarians were appointed Pope within a decade. Between 1047 and 1057 Damasus II, Leo IX and Viktor II (all Bavarians) were called to highest office in Rome.

Towns whose names end in -ing situated in Upper Bavaria and Austria – such as Pasing, Aibling, Uffing, Straubing, Efferding – are all places which were populated by the *Bajuwaren* between 550 and 1000 AD[5]. In 1007 the Roman-German King Henry II founded the bishopric of Bamberg. Two years later Henry again became Duke of Bavaria, which strengthened Bavaria's position when Henry was crowned German Emperor in Rome in 1014.

The House of Wittelsbach

In 1180, after the fall of Henry the Lion, Emperor Friederich Barbarossa transferred the dukedom of Bavaria to Bavarian Count Palatinate (*Pfalzgrafen*) Otto von Wittelsbach. This marked a turning point in Bavarian history. The dukedom of Bavaria became a state (*Territorialstaat*) – the dukedom of Austria had been surrendered in 1156 – and the House of Wittelsbach ruled Bavaria on and off for over 700 years. This was the beginning of a long period of artistic and cultural development, which reached a

---

[4] Freistaat Bayern. Eine kleine politische Landeskunde, published by the Bayerische Landeszentrale für politische Bildungsarbeit. 1989.
[5] Hans Rall. Zeittafel zur Geschichte Bayerns. Südd. Verlag. München. 1974. p.15.

climax under Ludwig the Bavarian (1302-47), who as German king and emperor acquired a lot of new territory, including Mark Brandenburg and the Tyrol.

Although Bavaria remained an agriculturally based area, residence and imperial cities developed under the Wittelsbach influence into intellectual, artistic and economic centres, some of which were even of European rank. In 1214 the Palatinate was acquired by the Wittelsbacher dynasty and therefore became Bavarian. The Palatinate was to remain connected with the fate of Bavaria's historical development over the next 700 years.

In 1255 the dukedom was first divided into Upper and Lower Bavaria (*Oberbayern, Niederbayern*), with Munich and Landshut as the respective capitals. In 1402 the Pope granted permission for Bavaria's first university to be founded. Only Prague, Vienna, Heidelberg, Cologne and Erfurt predate Würzburg as German universities. The university was closed, however, in 1411 when a student murdered the rector, Johann Zantfurt.[6] A new university was founded in Würzburg in 1582.

Before this, however, (in 1472) a university was established in Ingolstadt[7], by Duke Ludwig the Rich, with the support of Johannes Eck, a well known opponent of Martin Luther. The university of Ingolstadt was transferred, amidst some controversy, to the power of the Jesuits in 1542. Much later, in 1800, the university was moved to Landshut and then to Munich in 1826[8]. Close links between education and the Catholic church in Bavaria have always existed, and this has from time to time proved very controversial, even in the much more recent past.

The influence of the Reformation

Following the publication of Luther's ninety-five theses in 1517, the Reformation spread to Franconia and Swabia. The establishment of the Protestant (*evangelisch*) Church led to divisions into one or other of the two religious denominations. These splits became particularly noticeable during the

---

[6] Hans Nöhbauer. Die Chronik Bayerns. Chronik Verlag. Dortmund. 1987. p.136.

[7] Hans Dollinger. Bayern. 2000 Jahre in Bildern und Dokumenten. Bertelsmann. München. 1976. p.75 refers to die "Hohe Schul" zu Ingolstadt . . . "bis 1800 bayerische Landesuniversität."

[8] op.cit. Nöhbauer. p.620.

Counter Reformation under Albrecht V (1550-79) and Wilhelm the Pious (1579-97)[9]. These events were of fundamental importance in Bavaria, Franconia and Swabia, where religion, and more especially denomination and church attendance have always played, and even today continue to play, though to a lesser extent, a key role in both everyday life and political affiliation.

As will be seen later, the Catholic/Protestant dimension and links between the Church and political parties and their supporters were often the cause of much consternation in Bavaria during both the Weimar period and in the Federal Republic. The origins of such disputes have their roots in the eighth, twelfth and sixteenth centuries. The oldest document in Bavarian history in this connection is the instruction from Pope Gregory II in 716 for a plan of how to organise a church stucture for Bavaria. This plan was later put into action by Bonifatius (Saint Boniface) in 739, when he created bishops for Salzburg, Freising and Regensburg and confirmed Bishop Vivilo's position in Passau.[10]

When Bonifatius elevated Bavaria to a church province, this was seen as a political victory for the Agilofing rulers who wanted to create a counterbalance to Frankish power. The Bavarian historian Hans Rübesamen saw this as a struggle for independece from a superior and centralised power.[11] Rübesamen is not alone in seeing a strong desire for recognition as a separate, independent state with its own clearly defined identity and culture as a Leitmotiv of Bavarian history. This has endured right up to the present, still occurring in contemporary German politics.[12]

The enduring relevance of the religious factor

In an attempt to preempt any possible splintering of territory, an Act of Primogeniture, regulating the inheritance of titles and land, was passed in 1555. Its aim was to aid stability. Territorial division was to some extent avoided anyway by the fact that the Peasants War (1524/25) did not affect

---

[9] Information about Bavaria. Bavarian State Chancellery. Munich. 1987. p.51.

[10] Freistaat Bayern. Hanns-Seidel-Stifung. München. A publication which accompanied an exhibition on the Free State of Bavaria. No date given.

[11] Hans Eckart Rübesamen. Fourteen Hundred Years of Bavarian History, in Imago Bavariae. Herder. Munich. pp.185-193. No date.

[12] Ibid.

Bavaria to any great extent, although it caused considerable unrest in Franconia and Swabia.

Also in 1555 a "religious peace" was passed in Augsburg (*Augsburger Religionsfriede*). This was of great significance, because in law it – officially at least – brought to an end the disputes between Catholics and Protestants. Nine years after Luther's death the two denominations were granted equal recognition. At this time four fifths of the Germen population was Protestant. In fact the ideas of the Reformation did not take hold in Old Bavaria (*Altbayern*) to the same degree as for example in Franconia, which was, with the exception of the prince bishoprics, predominantly Lutheran; the focus of activity here was in the imperial city of Nuremberg and the Hohenzollern markgravia.

The religion and denomination of the inhabitants of a region was now determined by their ruler: *culus regio, eius religio*[13] (whose land, his religion). A further stipulation of the 1555 law meant that any clerical prince who converted to the Protestant faith automatically lost his office and land. The Dukedom of Bavaria contained only very small Protestant enclaves.

The first Concordat

In 1583 the first Concordat (*Konkordat*) was signed between Bavaria and Pope Gregory XIII on behalf of the Roman Catholic Church. This suited the Duke of Bavaria, William V, because he wanted to strengthen his links with the Catholic church. In fact he wanted to gain more land for his son, Ernest, who by 1585 administered over five bishoprics (Freising, Hildesheim, Lüttich, Münster and Cologne).

The Wittelsbacher dynasty, by gaining the diocese of Cologne, had in their possession until 1761 not only an important arch-bishopric but also the power and voting rights which went with it. This was to be the first, but not the only example of official treaties between Bavarian rulers and the Vatican. Such agreements were to play a crucial role in Church-State relations. Bavarian church history has been both significant and influential over many centuries.

The important role of the monasteries in religious, social and cultural life has extremely deep roots, with the result that Bavaria was sometimes

---

[13] Wessen Land – dessen Religion, on p.334 of Chronik der Deutschen. Brigitte Beier et all. Chronik Verlag. Dortmund. 1983.

called *terra benedictina*. Monasteries in Benediktbeuren and Tegernsee in the Alpine foreland, Weltenburg and Niederaltaich on the Danube, Andechs in Upper Bavaria, Füssen and Ottobeuren abbeys in Swabia and the Franconian monasteries of Münsterschwarzach and Amorbach still bear witness today to this strong Bavarian tradition.[14]

Bavaria becomes a Prince Electorate

During the period of the Thirty Years War (1618-48) Maximilian I, one of Bavaria's strongest rulers who re-structured administrative and financial arrangements, as well as creating a permanent army, played an important part in the Catholic League. This was founded in 1609 by Maximilian, as a reaction to a Protestant Union which had been established in 1608. The Union took as its aim the preservation of Protestant interests *vis-a-vis* the German Emperor in the Palatinate, Württemberg, Baden-Durlach, Ansbach-Bayreuth, Kulbach, Ulm, Straßburg and Nuremberg.[15]

The Thirty Years War brought great hardship to Franconia, Bavaria and Swabia. The situation improved, however, when in 1623, under Maximilian I, Bavaria became an Electorate (*Kürwürde*). The Upper Palatinate (*Oberpfalz*) came to Bavaria in 1628 and Maximilian became the first Bavarian prince/elector (*Kurfürst*). Maximilian I (1597-1651) proved himself to be a leading statesman and Bavaria reached a cultural zenith under his rule, charaterised by the Baroque and Rococo periods, which lasted until the reign of Max III Josef (1745-1777), who was the last elector from the Old Bavarian line of the Wittelsbach family.

Under Max Emmanuel (1679-1726) the Arts continued to flourish; unfortunately his attempts to pursue power politics on a grand scale failed. Owing to tension and rivalry with the Hapsburgs and Austrian interests, Bavaria was pushed towards becoming a natural ally of France. Prussia at that time was standing very much to one side and Austria was laying claim to Bavarian territory. Kurfürst Max IV Josef (1799-1825), who was from the Paltinate-Zweibrücken (to the left of the Rhine) Wittelsbach line, was forced, because of the difficult situation in which he found himself, to place

---

[14] op.cit. Freistaat Bayern. p.94/95.
[15] op.cit. Beier. p.340.

his land under the protection of Napoleon. Bavaria was about to enter a new political era.

Bavaria as a kingdom

On 1st January 1806 Bavaria became a kingdom (*Königreich*) by the grace of Napoleon and Max IV Josef became Max I Josef, Bavaria's first king (1806-1825). The kingdom of Bavaria united 83 fragments of territory from Lindau in the south west to Hof in the north east, and from Aschaffenburg in the north west to Berchtesgaden in the south east.

These territories were of varying size, origin and tradition. The new Bavaria (Neubayern) which came together between 1800 and 1816 resembled very closely the shape of present-day Bavaria, since it took in the territories of Franconia and Swabia. The main difference from the present-day situation was the Palatinate territory to the left of the Rhine (see fig. 1), which was a legacy from the Wittelsbach rulers, who had shaped Bavaria's destiny so significantly over some 700 years.

Bavaria, joined the Rhine Confederation (*Rheinbund*) and in 1815, with a population of 3.5 million and an area of 75,000 sq. km., was the third largest state in the German Confederation. The year 1806 marked the end of the Holy Roman Empire. Emperor Franz II had to renounce his crown, although he continued to rule as Franz I of Austria.

The modern state evolves

The chief architect of the modern state of Bavaria, however, was neither a king nor a prince but a leading minister: Maximilian Count (*Graf*) von Montgelas. He was born in Munich, although he was brought up and educated in the Savoy region of eastern France. Although considered more of a Frenchman than a German, Montgelas was undoubtedly a great Bavarian. He is justifiably often referred to as the father of the modern Bavarian state.

From 1799, as foreign minister, Montgelas, together with Max IV Josef, who was also something of a francophile, began to design and build the foundations for a new administrative and constitutional system in Bavaria. Between 1803 and 1806 Montgelas was both foreign and finance minister, and from 1806 onwards he was simultaneously minister for home affairs and education[17]. His dominance of Bavarian politics was complete. He brought in a new, tightly-run administrative machine, introduced compulsory schooling and the equality of the Christian denominations. He also abolished peasant surfdom.

In 1803 the Imperial Commitee (*Reichsdeputationshauptschluß*) awarded the bishoprics of Augsburg, Freising, Bamberg and Würzburg to Bavaria, as well as parts of the bishoprics of Eichstätt and Passau, thirteen imperial abbeys and fifteen imperial towns/cities in Franconia and Swabia[18]. Between 1803 and 1819 the territory of Bavaria doubled in size.

Together "Vater Max", as Bavaria's popular first king was affectionately known by his people, and his omnipotent minister, Montgelas, produced the first Bavarian constitution (*Konstitution*) in 1808. This was a fairly loose document, consisting of only six articles and 45 paragraphs[19], which attempted to establish broad principles only, especially in the area of justice and equality before the law.

The "revolution from above" certainly helped the Old Bavarians, the Franconians and the Swabians to evolve into a single people. Montgelas divided Bavaria at first into fifteeen, then thirteen, approximately equal regions, each named after a river on the French model. Soon afterwards the number of regions was reduced to eight, which in fact corresponded fairly closely to the seven administrative districts of present-day Bavaria (the eighth one was the Palatinate, which remained part of Bavaria until the creation of the new *Land* in 1946).

In 1818 a more detailed and effective constitution (*Verfassung*) was drawn up. This is generally regarded as Bavaria's first foundation for the establishment of a democratic parliamentiary system. The constitution was based on monarchical principles, and a Bavarian parlaiment (*Landtag*) was established, consisting of two chambers – a *Reichsrat* and an *Abgeordnetenkammer*. These constitutional arrangements proved to be successful in integrating the new territories which had joined the sovereign kingdom.

Again Church-State relations were not forgotten; in 1817 a new Concordat was signed with the Roman Catholic church, followed in 1818 by

---

[17] op.cit. Nöhbauer. p.283.

[18] ibid. p.283.

[19] op.cit. Nöhbauer. p.288.

an edict from the Protestant church. Both agreements were appended to the constitution. Although Montgelas, a master of delicate diplomacy, had been instrumental in laying the foundation stones of the modern Bavarian state, he was pushed aside and treated very shabbily (1817), partly as a result of disagreements with crown prince Ludwig during the absence of the king.

As King Ludwig I (1825-48), Maximilian's son presided over a liberal regime initially, influenced by the revolutionary events in France and Poland (1830) and especially the *Hambacher Fest* (1832), a massive demonstration by the German liberal movement in the Bavarian Palatinate. Bavaria joined the German Customs Union (1833) led by Prussia.

After a well publicised affair with a dancer, Lola Montez, Ludwig, also under growing political pressure from the March Revolution, abdicated in 1848. During his reign Munich had developed into one of Germany's outstanding cultural centres. Many of the present-day buildings and monuments in Munich bear witness to this. In Ludwig's day writers, artists, architects and scientists were summoned to the Bavarian capital from all over Germany.

The university was moved from Landshut to Munich in 1826, the monasteries were restored, a rail network developed – the first railway in Germany ran from Nuremburg to Fürth in 1835 – and many links with trade and industry were promoted during the time of Ludwig I. His son and successor Maximilian II (1848-64) continued as a patron of the arts; he also introduced social and political reforms (e.g. a new electoral law abolishing the electoral college), as well as furthering the cause of science. The kingdom was flourishing in an age of Romanticism.

Prussia or Austria?

Owing to the special nature of its idiosynchratic development, Bavaria has during the course of its long history often displayed the characteristics of a non-conformist. Bavaria and its inhabitants have from time to time rebelled against attempts to impose authority on them from external sources. This struggle to become and remain independent is a key theme which runs through Bavarian history. It has presented problems whenever Bavaria has been forced to make a decision on whom to side with in wars and disputes.

After the Thirty Years War, for instance, Bavaria had to manoeuvre skilfully to avoid being smothered by France and Austria, two very powerful neighbours. When Napoleon made Bavaria a kingdom, Bavaria at first fought with France, paying the price of the loss of 30,000 Bavarian lives when Napoleon invaded Russia. Shortly afterwards, however, Bavaria switched sides and joined the campaign against France in 1814. In this way Bavaria retained the territories it had gained from Napoleon long after he had been defeated. Similarly Bavaria sometimes sided with Austria, yet at other times fought against her.

Following the failed attempt to unite Germany in St Paul's church (*Paulskirche*) in Frankfurt in 1848, Prussia was determined not to give up. As matters came to a head in the power struggle between Prussia and Austria, Bavaria favoured the *großdeutsch* solution, since it expected greater room for manoeuvre from the inclusion of Austia.

Bavarian independence under threat

It was becoming increasingly difficult for Bavaria to pursue an independent policy in the second half of the 19th century. In the German war of 1866 Bavaria sided with Austria and lost. That defeat led to to the conclusion of a secret offensive and defensive alliance with Prussia. This opened the way for the *kleindeutsch* solution, and, when the German states were at last unified in 1871, Bavaria was part of the German Empire, not just led by, but completely dominated by Prussia and Bismarck.

Despite the hegemony of Prussia in the new empire, Bavaria was afforded special status. It was granted a type of "collective sovreignty."[20] Bavaria retained, in comparison to other states in the empire, a certain amount of limited (perhaps mainly psychological) independence via its special rights (*Reservatrechte*). These included military sovereignty (but only in times of peace), special rights in the Upper Chamber (*Bundesrat*) concerning foreign affairs, privileges in diplomatic relations abroad, independent Bavarian postal and rail systems and also the right to collect taxes on beer and spirits[21]. So yet again Bavaria, although she had by no means entered

---

[20] The phrase "Bundesstaat mit Kollektivsouveranität" is used by the Bavarian historian Karl Bosl.

[21] Rainer Roth. Freistaat Bayern. Politische Landeskunde. Bayerische Landeszentrale für politische Bildungsarbeit. München. 1992. p.48.

the Bismarck Reich full of enthusiasm, was able to claim a certain special status, behaving as an exception to the rule. These special rights were probably just as important to Bavaria in a psychological as in a practical sense.

The special position of Bavaria, though still a constituent member of the German Empire, is typical of her periodic "rebel" or "renogade" behaviour during the course of her long and rugged development. The Bavarians had once again made their point: although basically they had no choice in whether to join the new Empire or not, they insisted on some recognition of Bavaria's special position. Nevertheless important decisions were now taken in Berlin. Meanwhile in Munich the stage was being set for a most unusual Bavarian ruler.

The age of the "fairytale" king

In Upper Bavaria even today King Ludwig II (1864-86) is still referred to by some local inhabitants with great affection. You may well be told that he was not the "mad" king, although he was certified insane and removed from office in 1886. Bavarians prefer to remember Ludwig II as the fairytale king (*Märchenkönig*), who built the idyllic castles of Lindenhof, Herrenchiemsee (modelled on Versailles), Hohenschwangau and perhaps the one most visitors know best, Neuschwanstein, seen on so many posters and often associated by tourists with Bavaria's romantic holiday image.

Ludwig II was young, handsome, frequently melancholy, seeming to live in a world of his own – a dream world which apparently offered him an escape from reality. Although his reign is recalled fondly by many Bavarians and pictures of him still appear on Bavarian postcards even today, Ludwig was, paradoxically, completely out of touch with his people and their everyday lives. He was much more interested in architecture, art and Wagner's music (the opera house in Bayreuth was opened in 1876) than politics or the state's finances.

Owing to his total neglect of his official duties as Bavarian Head of State and the increasingly serious financial situation, there was growing pressure to replace Ludwig. He was declared mad (mentally ill) and therefore unfit to rule, by doctors who did not examine him, and taken to Berg castle as a prisoner. Two days later he was allowed out for a walk around

Lake Starnberg, together with his chief physician Dr von Gudden. Both men were later found drowning in the lake.

The speculation and rumour concerning precisely what happened have, to this day, never been completely clarified. Either suicide or an escape attempt were suspected but never proved. Shortly before these events, which led to his death on 13th June 1886, Ludwig had been informed that, since he was no longer considered capable of managing the affairs of state, and since the next in line to the throne, his brother Otto, had been mentally ill for years, Ludwig's uncle, Prince Luitpold, would take over as Prince Regent (*Reichsverweser*)[22].

Regency rule in Bavaria

As Luitpold (1886-1912) was not a king, his constitutional powers were limited. Bavarian history books record that officially Otto I was the king of Bavaria from 1886 until 1913 and Luitpold, and later his son Ludwig, ruled as Prince Regent in his place. Although foreign policy was controlled by Bismarck anyway, there was actually considerable progress in educational and cultural matters in Bavaria under the new Prince Regent. Most of the work was undertaken not by Luitpold himself, who was aged 65 when he took over, but by his ministers, the most prominent of whom was Lutz.

More schools were built than ever before, and the universities of Munich, Würzburg and Erlangen were expanded, as well as the *Technische Hochschule* in Munich, considered the best institution of its kind in the Reich in 1910. The *Deutsches Museum*, developed gradually from 1906 onwards, and the *Nationalmuseum* was also built.[23]

Poets, painters and writers gathered in the fashionable artists' quarter of Schwabing where Munich's "high society" gathered. Thomas Mann, who moved from his native Schleswig-Holstein to settle in the Bavarian capital, is reported to have said that Munich was "ablaze with light" at that time[24]. Technological advance and industrialisation were accompanied by population growth in Bavaria's major cities (Munich, Nuremberg) and the development of centres of heavy industry (Schweinfurt, Selb).

---

[22] op.cit. Nöhbauer. p.371.

[23] op.cit. Dollinger. p.178.

[24] op.cit. Rübesamen. p.193.

Luitpold, a passionate devotee of mountaining and hunting and a real "man of the people" became a symbolic figure for the kingdom of Bavaria, even though he was not actually a king. The reason for this was that his life spanned almost the whole epoch from Montgelas and Bavaria's first king, King Max I Josef, to its last monarch. Despite Luitpold's popularity and the achievements of his ministerial bureaucracy, in political terms it was a time of stagnation. Prussian hegemony was continually in evidence and the *Kulturkampf*, whereby Bismarck made repeated attempts to stem the influence of the Catholic Church, had restricted Luitpold's room for manoeuvre almost before his rule began.

When Luitpold died in 1912, his son, Ludwig succeeded him, initially also as Prince Regent. However he really wanted the more powerful position of king and in 1913 he proclaimed himself King Ludwig III. This brought about an awkward situation because the "nominal" king, Otto, the mentally ill brother of Ludwig II, was still alive (he died in 1916). The standing of the Bavarian monarchy suffered greatly.

The outbreak of the first world war aggravated an already deteriorating political situation. The call for greater responsibility of the ministers vis-a-vis the elected parliament was long overdue. When this demand finally came from the SPD in September 1917, it was initially rejected; when the concession was at last granted in November 1918, it was too late. By then the Bavarian monarchy could no longer be sustained. Luitpold and Ludwig III, the last King of Bavaria, were the last representatives of the House of Wittelsbach, which had ruled Bavaria for 738 years. It was the end of an extremely important era in Bavarian history. Radical changes were ahead in what was to be an explosive and revolutionary period for Germany in general and Bavaria in particular.

The short-lived Socialist Republic

The truly staggering and unexpected events which shaped Bavarian politics towards the end of the first world war, completely unforeseen by anyone, were to have a lasting effect. Like so many other people, the Bavarians were desperately awaiting the end of an appalling war which had produced so much death and destruction.

Early in 1918 the SPD split, and a minority breakaway group of independent socialists (USPD) was formed under the Berlin socialist Kurt Eisner. Eisner, a revolutionary idealist and journalist who had been the editor of the SPD newspaper *Vorwärts* until 1905[25], tried to extend the wave of strikes against the war. These had begun with a strike of some 42,000 munition workers in Nuremberg in January 1918.

In the autumn of that year Eisner formed a provisional workers' and soldiers' council and proclaimed Bavaria to be a "Free People's State" (*Freier Volksstaat Bayern*). Amidst cries of "down with the king", Eisner got himself elected Bavarian minister president by his council. On the same day (7th Nov.) King Ludwig III fled from Munich. Bavaria was declared a *Freistaat* (republic) – no longer a monarchy. This was the first time that the expression *Freistaat Bayern* – the Free State of Bavaria – was used. It is interesting to note that even on this unusual Bavarian council of soldiers and workers there was a representative of Bavaria's agricultural interests, Gandorfer, who was a member of a well known farming family from Lower Bavaria. The importance of agriculture in Bavarian society has deep roots and an enduring presence, still evident today.

Eisner's new regime was unable to sustain support either in parliament or amongst the people, as hunger and economic mismanagement prevailed, and at the first elections for the Bavarian parliament a few months later, in February 1919, Eisner's party, the USPD, gained only three mandates.

As he was walking to parliament to hand in his resignation on 21st February 1919, Eisner was shot dead by Count Anton von Arco-Valley, who was immediately celebrated as a hero by the Conservatives. The assassin was tried and initially given a death sentence, which was subsequently commuted to life imprisonment. Some years later a full pardon was granted, and Arco-Valley became a speaker for the Bavarian People's Party (BVP).

Following the murder, street demonstrations and violent riots broke out and a socialist republic (*Räterepublik*), promoted by Communists, was proclaimed in early April. The "red revolution" forced the government to flee from Munich to Bamberg. For a brief period chaos and virtual anarchy reigned. After much bloodshed the so-called "white" Guard, consisting of soldiers from Prussia, Bavaria and Württemberg, ended the "red" Republic

---

[25] op.cit. Nöhbauer. p.437.

at the beginning of May, just after Lenin had sent a telegram advising the revolutionary forces on how to organise themselves against the "bourgeois henchmen"[26]. The revolution had spread to Munich university in April 1919, where a revolutionary council had briefly replaced the academic Senate. Discussions on the Communist Manifesto of Marx and Engels took place; an experimental proletarian university was established but abandoned after only two weeks.

Bavaria had been shaken to its very foundations. The depressing experience of the Great War, in which over nine million had died, the Russian Revolution of 1917 and her own personal experience of a socialist republic in Bavaria would never be forgotten. All these factors, especially the memory of the *Räterepublik*, played a key role in shaping Bavarian politics in the 1920s and 1930s. The political Left had been thoroughly disgraced in the eyes of most Bavarians, who now harboured a deep distrust of all things Socialist or Communist. Politics in Bavaria would never be the same again.

Germany's first democracy – the Weimar Republic

August 1919 saw the start of Germany's first attempt to establish a parliamentary democracy. Earlier that year two new political parties had been founded: the German Communist Party (KPD) in January, and the German Workers' Party in the same month. The latter was founded in Munich by Anton Drexler. In the autumn of 1919 a man named Adolf Hitler joined the party, which in February 1920 was re-named the National Socialist German Workers' Party (NSDAP).

Bavaria, within the framework of the new German Republic, received her first republican constitution, known as the Bamberg constitution, named after the temporary domicile of the Bavarian government. Bavaria, which gained the territory of Coburg after a referendum in 1920, was described as a Free State (*Freistaat*) or republic and member of the Reich.

Pressure from the Right

It was now the turn of the political Right to threaten the young German

---

[26] op.cit. Nöhbauer. p.445.

democracy. Right-wing extremists, monarchists and a collection of anti-communist and anti-semitic forces began to assert themselves. The Kapp Putsch, which took place in Berlin in March 1920, was led by Wolfgang Kapp from East Prussia. He declared himself Reichskanzler after his right-wing troops had taken over the National Assembly.

Although the coup collapsed after only four days, it affected Bavaria too, where SPD minister president Hoffmann resigned. There was a general strike in Nuremberg, called by the trade unions reacting to the attack from the Right. Bloody clashes on the streets followed. Von Kahr, who headed the Bavarian government after Hoffmann, was a declared monarchist who wanted to reinstate the pre-1918 status quo. At the very least he and his supporters wished to create a nucleus of order in Bavaria (*Ordnungszelle Bayern*), regardless of what was happening in the rest of Germany. Tensions between Bavaria and the Reich developed. The atmosphere was explosive. The fuse merely needed to be lit.

The spark came in November 1923 with the Hitler Putsch. Adolf Hitler called for a national revolution and declared the German government to have been dismissed. Initially von Kahr tolerated the attempted coup by Hitler and his supporters (around 2,000 armed National Socialists). Hitler wanted to march to Berlin, carrying the flag with the old colours of the Reich – black, white and red – but only got as far as the Feldherrnhalle monument in Munich on the next day, which was the 9th Nov. 1923, five years to the day since Germany's collapse.

After some hesitation von Kahr eventually acted, the Bavarian police opened fire and the putsch ended in a fiasco. In Feb. 1924 v. Kahr resigned. (Hitler had him murdered in 1934 as part of the "Röhm Putsch")[27]. General Erich Ludendorff was pardoned for his part in the attempted coup. Hitler and three others – Weber, Kriebel and Pöhner – were sentenced to five years in prison (five others received fifteen months each)[28]. Hitler was, however, released from prison in Landsberg after only nine months. After the trial the court issued a statement in which it was apparent that some of the judges were sympathetic to Hitler's actions. Hitler was described as a highly talented man and a gifted speaker who had fulfilled his duty as a

---

[27] op.cit. Roth. p.51.
[28] op.cit. Dollinger. p.214.

soldier. He was not to be reproached for his great enthusiasm for the German cause (die deutsche Sache), which had been the chief motivation, rather than simply personal ambition. Hitler, it was stated, had worked his way up from a very humble background to a respectable position by means of serious and hard work[29]. This could hardly be taken as a condemnation of the Hitler Putsch.

The exceptionally mild treatment by the courts of this act of high treason has been the object of a great deal of research, especially against the background of the later recovery of the NSDAP from around 1929/30 onwards. Clearly the total collapse after a world war, the end of what for many was a golden age of the Wittelsbacher dynasty associated with the kingdom of Bavaria, the deeply socialised fear of the Soviets after the communist revolution in 1917, and the negative experience of the socialist republics set up in various parts of Germany (particularly the one in Munich) played a major role in determining how Bavaria reacted to the emergence of right-wing political forces.

In addition it should not be forgotten, of course, that many Germans felt deep resentment at the crippling conditions imposed by the Treaty of Versailles (reparations) and Germany's low status after the igmony of defeat. Furthermore, in September 1923 galopping inflation meant that 100 million mark bank notes were being printed. By June 1924 a litre of milk cost 1440 marks and a loaf of bread 2,500 marks. In the context of the prevailing circumstances in Bavaria and Germany at the time, and given what had occurred between 1914 and 1923, there were grounds for thinking that not everyone disapproved totally of what Hitler had attempted to do.

Right-wing forces gather strength

As Hitler started his prison sentence in April 1924, elections to the Bavarian parliament were held., in which the *Völkischer Block*, a successor to the banned NSDAP, gained just over 17 per cent of the vote and 23 seats at its first attempt. Bavarian politics at that time was experiencing a clear shift towards the right-of-centre parties.

After his premature release from prison, during which time he had writ-

---

[29] op.cit. Beier at al. p.809. "Hitler as a martyr." Also quoted from "der Strafantrag," issued by the chief prosecuter Ludwig Stenglein on 21.03.1924.

ten *Mein Kampf*, Hitler re-established the NSDAP in February 1925. The Bavarian government was headed by Held, a convinced federalist, from the largest party, the Bavarian People's Party (BVP). In contrast to the centralised Weimar constitution, Bavaria was able to re-gain a little of her traditional independent spirit for a brief period. Shortly before the elections, in March 1924, the Bavarian minister president von Knilling and the papal nuncio in Munich, Pacelli, who later became Pope Pius XII, signed a Concordat between the Bavarian State and the Catholic Church. The *Konkordat* regulated relations between the two partners, emphasising the Christian nature of Bavarian society and the right of the Catholic Church to lay down guide-lines on schooling and, in particular, on religious instruction.

At the same time the Bavarian government signed treaties with the Protestant Church to the right of the Rhine, as well as with the united Protestant-Christian (*protestantisch-evangelisch-christlich*) Church in the Palatinate (to the left of the Rhine). It was not until 1933 that a Concordat was signed between the Catholic Church and the German Reich.

In 1926 the much-loved Munich satirical magazine, Simplicissimus, well known for its hard-hitting cartoons caricaturing political figures and events of the day, celebrated its 30th birthday. Another popular Munich weekly, Jugend, was also established in 1896 as an artistic fashion magazine. The illustrated weekly was sold to another publisher in 1926, after the original publisher, Hirth, ran into financial problems.

Progress in science and technology continued, as the demand for private cars took off. In spite of the general economic crisis, the Bavarian economy, dependent mainly on agriculture and small and medium-sized businesses, was not as badly affected as many other German states. The withdrawal of American credits affected mainly the big concerns, of which Bavaria at that time had few. At the third NSDAP party conference, held in Nuremberg in August 1927, Hitler made his first public speech for two years. A ban which had been imposed on him by the Bavarian government two years earlier was lifted in March 1927.

National Socialism marches forward

After the success of the *Völkischer Block* party in the 1924 elections to the Bavarian parliament, the NSDAP, re-established in 1925, fought its first

41

election to the *Landtag* in Munich in 1928 and gained only 6.3 per cent. At the next elections four years later the National Socialists polled 32.5 per cent. The BVP remained the largest parliament in Bavaria, but only just.

Meanwhile the NSDAP, assisted of course by the SS and SA and the various terror campaigns, gradually increased its share of the vote in the *Reichstag* in Berlin. Hitler's party increased its vote from 2.6 per cent in 1928 to 18.3 per cent in 1930, making it the second largest party, to 37.4 per cent in July 1932, by which time mass unemployment in Germany had reached six million.

In July 1932 an organisation called Bavarian Watch (*Bayernwacht*) was formed by Christain trade unions, Catholic associations and farmers' organisations in Bavaria. This was at the suggestion of Fritz Schäffner, the Bavarian finance minister and chairman of the BVP. The *Bayernwacht* was a uniformed organisation set up for the protection of its own interests against the SA and SS. In 1933 it was banned by Hitler.

On 15th March 1933, with Hitler already firmly established in power, Held and the Bavarian government were forced to resign. With the disbanding of the Länder (*Gleichschaltung*) in late April/early May 1933, Bavaria ceased to be an independent political unit until it was re-established in 1945/46.

Many Bavarians resent the fact that Bavaria was, and still is, associated in many people's minds with the National Socialist dictatorship. Munich was seen as the capital of the NS "movement" (*Hauptstadt der Bewegung*). Nuremberg was the city of so many of Hitler's rallies and its name was also associated with the Nazi race laws and later the trials. The first concentration camp was in Dachau. For months before the *Reichskristallnacht* synagogues were destroyed in Munich and Nuremberg. Quite apart from the connections between Hitler (an Austrian) and Munich, there were other big names in National Socialism who were Bavarian: Ernst Röhm (SA), Julius Streicher (*der Stürmer*), Hermann Göring and Heinrich Himmler.

To counter-balance this negative image, it should of course not be forgotten that there was also tremendous courage displayed in resistance movements in Bavaria, as in other parts of Germany. Perhaps the most notable one in Bavaria was the daring White Rose (*die Weiße Rose*) group led by the students Hans and Sophie Scholl and supported by Professor

Kurt Huber, Christoph Probst and Alexander Schmorell[30]. Following total collapse and the end of the second World War on 8 May 1945, it was not only Bavaria which was waiting for a new political system.

---

[30] These points are referred to in op.cit. Roth p.52. The author gained the impression in many interviews and discussions in Bavaria that this association of the Third Reich with Bavaria in general and Munich in particular is resented by some Bavarians.

3

The Postwar Bavarian Political System

The *Land* of Bavaria, which came into existence in December 1946, was sub-divided into seven administrative districts. These corresponded very closely to those established by Count Montgelas shortly after Bavaria became a kingdom in 1806, thus ensuring continuity and respecting the traditions of the various regions of Bavaria.

Originally Montgelas, influenced by French administrative models, divided Bavaria into fifteen districts, each named after rivers (Main, Danube, Lech, Isar, Pegnitz, Regen etc)[1]. From 1817 onwards the number of districts was reduced to eight. At that time Bavaria possessed an eighth distict in the Palatinate, which included Speyer to the left of the Rhine. When this territory was lost after 1945, the remaining seven districts were reestablished. In this way, not only was Bavaria's thousand-year history reflected, but the different regions with their own special and enduring historical and political traditions were also preserved.

---

[1] See chapter one. The 15 districts were devised in 1808, adapted to 8, based on the rivers Rhine (Palatinate), upper and lower Main, Rezat, Regen, upper and lower Danube and Isar in 1817. Under King Ludwig I the districts received their current names. Nöhbauer. Die Chronik Bayerns. Chronik Verlag. Dortmund. 1987. p.296.

The seven administrative districts are re-instated

These seven regions of modern Bavaria, described by Mintzel as representing between them the three historical and political traditional zones[2] of modern Bavaria, were established originally during the period 1806-20. It was then that Bavaria gained the territories of Franconia (consisting of three districts, Upper, Central and Lower Franconia) and Swabia. The region of Swabia straddles part of Bavaria and part of Baden-Württemberg.

The oldest and most traditional part of Bavaria (see chapter two) is Old Bavaria (*Altbayern*), which consists of the three districts of Upper and Lower Bavaria and the Upper Palatinate. Present-day Bavaria consists of the seven administrative districts (*Regierungsbezirke*) shown on the map.

The Bavarian state government today continues to see itself as having a European role. This has been part of party political programmes for many years. The CSU, for example, projects itself as a Bavarian, a German and a European party. This was again confirmed in the party's fifth programme of principles (*Grundsatzprogramm*) – the previous one had been in 1976 – passed at the Party Conference in Munich on 8/9 Oct. 1993.

Under the heading "Unity in Diversity" (*Einheit in Vielfalt*) the programme speaks of European cooperation in a community in which national diversity remains. It emphasises the commitment of Bavaria, the oldest state in Europe, and the CSU to the European unification process.[3]

It is interesting to note that Napoleon laid a foundation stone in the early nineteenth century in Bavaria to mark the geographical centre of Europe. The Napoleon Stone was laid on a piece of high ground, the Tillenberg, in the present-day community of Neualbenreuth, near Waldsassen in the Upper Palatinate, and a granite rock, close to the original stone, can still be seen today, marking "the centre of Europe."[4] This is perhaps another of the psy-

---

[2] ... schließt das heutige Bayern drei große historisch-politische Traditionszonen ein, die fränkische, die schwäbische und die altbayerische. Alf Mintzel. Geschichte der CSU. Ein Überblick. Westdeutscher Verlag. Opladen. 1977. p.29.

[3] Grundsatzprogramm der Christlich-Sozialen Union in Bayern. Atwerb Verlag. München. 1993. p.121.

[4] This is referred to in Alois Fink. Bayern in Europa. München. 1965, and a picture of the "Granitsäule" on the Tillenberg near Waldsassen is shown on p.9 of a publication by the Bayerische Landeszentrale für politische Bildungsarbeit. Freistaat Bayern. Politische Landeskunde, by Rainer JA Roth. Munich 1992.

Fig 3.1 The three traditional zones (heavy lines) of Old Bavaria, Franconia and Swabia and the seven administrative districts (*Regierungsbezirke*), with their capitals, of the modern state

Taken from:
A Mintzel, Die CSU 1945-72.
Westdeutscher Verlag.
Oplanden. 1975. p.61

chological factors which, regardless of their practical significance, have had and, for some Bavarians, still have an intangible meaning.

For nearly fifty years Bavaria possessed a 775 km. long border along the "iron curtain". With the new developments since 1990 in eastern Europe, Bavaria now feels, more than ever, that it has a vital role to play at the heart of Europe, not only in the European Union, but also as a go-between in East-West relations.

One of the many interesting paradoxes in Bavaria is to be found in the unusual mixture of its backward-looking, nostalgic emphasis on past traditions, whilst at the same time displaying great pride in modern technology and a forward-looking approach to future progress in a European context. It is important to remember, however, that the seven administrative districts of modern Bavaria vary in their individual make-up. Some parts are more Bavarian than others.

The seven administrative regions of Bavaria

Upper Bavaria

This is the largest of the seven districts with one quarter of Bavaria's area and the greatest population density, and with nearly four million inhabitants[5]. Munich is the seat of government for this district. Obviously Munich, as the Bavarian capital, has played a major part in making Upper Bavaria a centre of cultural, economic, scientific and artistic life. Munich has been called a "village with a million inhabitants" (*Millionendorf*), and at the same time a "metropolis with heart" (*Weltstadt mit Herz*). No wonder that the Bavarian capital, which today still displays so proudly so many reminders of Bavaria's glorious history, was often called Germany's "secret capital" (*die heimliche Hauptstadt*).

Ingolstadt has developed into a modern industrial centre for cars (Audi) and refineries, despite the dominance of Munich, with Siemens, BMW (*Bayerische Motorenwerke*) and the many micro-chip related firms.

---

[5] All population figures are taken from combining the 1991 official figures, given in op.cit. Roth, with the additional 1993 numbers. Bavaria's population has increased steadily over the years. It is still increasing, owing to two factors. Firstly, the excess of births over deaths and secondly, the influx of foreigners, especially from the East in recent years.

Upper Bavaria has many tourist attractions: the Alps, wonderful scenery and beautiful lakes, monasteries such as Ettal, Benediktbeuren and Frauenchiemsee, and fortresses and castles such as Burghausen, Linderhof, and Neuschwanstein.

Lower Bavaria

The administrative district of Lower Bavaria, although often in the shadow of Upper Bavaria, represents an even older region. Only just over a million people (just under ten per cent of Bavaria's total) live in this district, which covers around 15 per cent of Bavaria's total area. The regional capital is Landshut, whose cathedral has the tallest brick spire in the world, whilst the cathedral in Passau boasts the biggest church organ in the world.

Lower Bavaria was typical of the radical change Bavaria went through after the second world war, changing its image from that of a predominantly agricultural area to one open to tourism, especially in the Bavarian Forest and economic expansion, e.g. the "free trade harbour" on the Danube at Deggendorf. The opening of Passau University in 1978 also boosted the region's image.

The old cultural city of Landshut was the centre of Bayern-Landshut, one of the three Bavarian territories in 1392. Passau, a traditional seat of the Catholic bishops dating back to Roman times, has also been instrumental in shaping Lower Bavaria's development. Towns such as Kelheim, Straubing, Deggendorf Zwiesel and the monasteries of Niederalteich and Metten, as well as the numerous Baroque churches remind the visitor of the area's strong cultural heritage.

Upper Palatinate

The third district of Old Bavaria is the Upper Palatinate, whose regional capital is Regensburg, one of the oldest cities in Bavaria. In terms of ancient tradition and history as a bishopric and free city of the Empire, Regensburg, which was for a time the capital of the German Reich, also dates back to the time when the Romans built a bridge over the river Regen and founded Castra Regina. This is the least densely populated district, containing nearly fourteen per cent of Bavaria's surface area but only nine per cent of its population – about one million. Initially this Bavarian heartland suffered from a lack of resources and poor communications, being rather cut off geographically, bordering on the former Czechoslovakia.

Gradually, the tertiary sector, for example Regensburg University, and new industry in the form of cars and electronics were developed. When the long-running Wackersdorf controversy over the recycling and burning of atomic waste was finally settled, considerable progress was made in the regional economy, which previously had been based almost exclusively on the coal and steel industry around Sulzbach-Rosenberg. Nevertheless there are still problems of restricted economic development in some parts of the region, for instance in the west (Neumarkt), the north (Tischenreuth) and the east (Cham), where small, traditional branches of industry still dominate the remote rural scene.

The recent relaxation of what were formerly very restrictive eastern borders is already beginning to have an impact on economic revival in the Upper Palatinate, an area which has always been more reknowned for its century-old historical festivals, charming customs and cultural monuments, e.g. the Walhalla near Kelheim. It is here, close to Waldsassen and the *Napoleonstein*, that the local inhabitants hope that the area which marks "the centre of Europe" in one of Europe's oldest states, Bavaria, will be able to participate in a Europe-led economic recovery.

Lower Franconia

Approximately eleven per cent of Bavaria's inhabitants, some 1.26 million, live in the administrative district of Lower Franconia, which accounts for about 12 per cent of Bavaria's total territory. This is the only one of the three districts of Franconia whose population is predominantly Catholic.

The lower Main region based on Aschaffenburg in the extreme north west of Bavaria has links with the neighbouring state of Hesse and, in particular, the business and banking centre of Frankfurt; it is, therefore, a region of some economic potential. It includes Würzburg, the regional capital, which is a bishopric and university city, as well as being a centre of wine production. Good communications and infrastructure are a positive feature here. The Baroque churches and the famous castle (*Residenz*) contain innumerable art treasures. Schweinfurt is an important industrial centre (Sachs, Fischer ballbearings). Bad Kissingen is a modern health spa resort.

Upper Franconia

Bayreuth is the regional capital of the predominanatly Protestant administrative district of Upper Franconia, which occupies just over one tenth of
Bavarian soil and contains slightly more than nine per cent of its population, just over a million. This region originally suffered more than any
other in Bavaria from its partially remote position next to the *Zonengrenze* –
the border with the former GDR.

Communication links were neglected over a long period and the
region also suffered from the lack of a big city with a back-up support network to aid sales and marketing. This meant that several centres, rather
than one main one, developed: Bayreuth, Bamberg, Hof, Coburg. It is
hoped that this district, following the re-unification of Germany, will now
be able to regain the pivotal function it used to possess under the Staufer,
Hapsburg and Hohenzollern rulers in the past.

There are already joint ventures between Bavaria and the federal states
of Saxony and Thuringia, some of which should assist the Upper
Franconian economy.

Franconian "Switzerland" is one of Bavaria's most picturesque tourist
attractions. Bamberg, an arch bishopric with a celebrated cathedral and a
thousand-year history, was, a few years ago, voted a "dream town" by the
Germans themselves. Since 1979 Bamberg once again has a university.
Bayreuth, the district capital, also has a university, but is of course most
famous for its annual Wagner music festival. Coburg joined Bavaria in
1920. It is a Protestant enclave which once offered Martin Luther refuge in
its famous fortress (*Veste*).

Central Franconia

Approximately forteen per cent of the population of Bavaria, around 1.6
million people, live in this district, which covers a little more than 10 per
cent of Bavaria's territory. Although Nuremberg, with a population of
almost half a million, is Bavaria's second city, the regional capital of
Central Franconia is the traditional markgrave town of Ansbach.

This administrative district is densely populated and extremely pro-
ductive. There are two sides to it: the highly industrialised urban conurba-
tion of Nürnberg-Fürth-Erlangen-Schwabach and the modern leisure and

tourist developments in the west of this region. The Rhine-Main-Danube canal was just one of several structural improvements in the region.

The enduring traditions and pride felt by today's inhabitants stretch back to the Middle Ages. However this sense of tradition and the spirit of former imperial cities like Nuremberg blend in with the modern industrial age. Nuremberg, with MAN (*Maschinenfabrik* Augsburg/Nürnberg) and many other industries, is well known too for its museums (e.g. Germanisches Nationalmuseum), churches – for example the Lorenzkirche – and famous Christmas market.

Rothenburg and Dinkelsbühl are still preserved like "jewels" from the MiddleAges. By way of contrast, the little town of Hilpolstein near the university town of Erlangen is the point from which the Rhine-Main-Danube canal connection was made in 1992 to form the link between the North Sea and the Black Sea.

Swabia

Bavaria's third city is Augsburg, which is the seat of government in the administrative district of Swabia. Augsburg, a Catholic bishopric and university city with a 2,000-year history, was a free imperial city, first founded as a Roman settlement. In the Middle Ages, the wealthy Fugger family, who were bankers and merchants, established the *Fuggerei*, in 1514. This can still be seen in Augsburg today. It was an estate of 67 houses for the city's poor citizens. At that time Augsburg was a very important trading centre.

In terms of scenery, Bavarian Swabia – Swabia straddles the federal states of Bavaria and Baden-Württemberg – contains some very attractive areas, for example Neu-Ulm, the Allgäu (around Kempten), as well as Augsburg itself. Cultural and artistic connections abound, although Augsburg, situated only 60 km. (40 miles) from Munich was somewhat overshadowed by the Bavarian capital during the 19th and 20th centuries. The idyllic "island of flowers" (*Blumeninsel*) Lindau, in the extreme southwest of the Free State, attracts enormous numbers of visitors in the summer. It was lost to Bavaria after 1945 but returned in 1956.

Swabia has an excellent balance between agriculture, farming and artisan occupations on the one hand and modern industry on the other. The engineering giant MAN is based in Augsburg, as well as in Nuremberg.

The Bavarian Institute for Waste Disposal is to be found in the Swabian capital and Memmingen is a centre for applied micro electronics. Banking, insurance and the tourist industry are also important.

Three administrative tiers

As we have seen, for administrative purposes Bavaria consists of seven districts – *Regierungsbezirke*, each with its own elected regional government in its regional capital. This is the third tier or level of administration (*dritte kommunale Ebene*). Each of the seven districts is sub-divided into rural counties (*Landkreise*), of which there are 71 in the whole of Bavaria, and county boroughs (*kreisfreie Städte*), of which there are 25 altogether. This represents the second tier of administration (*zweite kommunale Ebene*). The first tier (*erste kommunale Ebene*) are the municipalities, or communes (*Gemeinden*), the smallest administrative unit, of which Bavaria has 2052. About 80 per cent of the communes have less than 5,000 inhabitants.

Bavaria also has eighteen regional planning associations, which between them cover the the whole state and work together with neighbouring states, including some of the new Länder, in order to ensure that forward planning follows a logical and uniform pattern.

Political representation

Bavaria has a three-tier system. Below the state/ministsry level – the *Land* government in Munich – each of the seven administrative districts is headed by a president, who is an appointed civil servant. The electorate elect a district parliament (*Bezirkstag*) for each of the seven districts, which concerns itself mainly with administrative matters.

The electorate of the counties elect a president (*Landrat*) and a council (*Kreistag*) for each of the 71 counties, plus town councils and a mayor for each of the 25 county boroughs. The lowest administrative level is the town or village council (*Gemeinderat*).

A look back through history reveals that the strong emphasis on "grassroots" democracy and the desire to involve the citizen as much as possible at every level in the democratic process, evident in Bavaria today, are a legacy from the early 19th century. Territorial changes in 1808 and 1818 re-organised approximately 40,000 towns, villages and hamlets into around 7,000 communes. The decree (*Gemeinde-Edikt*) of 1818 granted each commune, as a political unit, a degree of autonomy in matters of administration.

It was not until 1869, however, that the parameters of responsibility were more closely defined. After the communes or municipalities were brought under the centralist leadership of the National Socialists in 1935 by the *Deutsche Gemeindeordnung* a new system was created in 1952.

Twenty years later the number of communes in Bavaria was reduced in a fundamental re-organisation of territory (*Gebietsreform*), which included reducing the number of *Kreise* from 143 to 71 and the 48 *kreis-freie Städte* to 25. A radical reform, which later also re-distributed and re-organised the allocation of agricultural land (*Flurbereinigung*) was carried through, the nucleus of which concentrated on strengthening the position of the *Gemeinde*, particularly the smaller ones with between one and five thousand inhabitants. These far-reaching reforms meant that such communes, as well as the larger administrative units, were able to gain a high degree of autonomy in decision-making on issues which affected their every-day lives. Bavaria's whole administrative structure became more transparent.

The executive

Bavaria is the only federal state in Germany where the state government consists of the minister president, the nine ministers *and* their secretaries of state. The minister president is supported by the Bavarian State Chancellery (*Staatskanzlei*), according to art.52 of the constitution.

The Chancellery is not a ministry or department. Its function is to assist the Bavarian premier in the numerous coordination and support tasks. Staff work in twelve departments, including press, personell, protocol, data protection, the organistion of exhibitions on Bavarian history etc. These 330 staff used to be housed in seven separate buildings. The Chancellery is where the Executive is located; the Bavarian cabinet holds its regular Tuesday meetings there.

The actual building has been the subject of much controversy for a number of years now. Even in the early 1980s, in the days of Franz Josef Strauß, the idea was generally considered to be unnecessarily pretentious and much too expensive by the "man and woman on the street" in Munich. A new building was constructed around an ancient Greek-style "temple," with a cupola on top, on the site of the former army museum founded in 1879 by King Ludwig I (building on the site actually dates back to 1560 under Duke Albrecht). In Bavaria the desire to build upon past legacies is never far away.

The Chancellery itself is situated on the edge of the famous Hofgarten park, whose history stretches back to the 16th and 17th centuries. It was originally christened the "Straußoleum" by the local inhabitants, owing to its similarity to a mausoleum. Even under Strauß, Bavaria's most popular premier, the project was criticised. Although the decision to build was taken in 1961, it was not until 1989, under minister president Streibl, that the plans were finally put into practice.

The new building was extended and finally finished in May 1993, amidst even more controversy, because of its size, since it is larger than the White House in Washington, and its rather pompous, avant-garde design. It has been given many more facetious names since its latest "face lift," including the Bavarian Kremlin, and later the Amigodrome, after the original "Amigo" affair which resulted in the resignation of Max Streibl; the Bavarian Greens called it a "swanky" imperial building.[7]

The dome and Wittelsbach statue in the centre are visible from the rear only, and unfortunately this "classical" architectural style clashes with the rest of the modern glass structure. It is apparently felt by many of the Munich people, according to radio phone-in programmes and press surveys, that the excess of glass makes it look like a cross between a massive greenhouse and a modern shopping centre. The astronomical cost to the Bavarian state of DM 241.8 million was, for many, the final straw.[8]

In spite of such objections, however, a small section of the civil servants and politicians who work there seem to think the mere existence of the *Staatskanzlei* in its fine setting of the Hofgarten lends authority to the Bavarian state government in Munich.

---

[7] Spiegel 19/93, p.108.

[8] ibid,. p.108. Spiegel reports the cost as DM242 Million (around £97m or $142m). The cost is stated as DM241.8m in an information brochure "die Bayerische Staatskanzlei."

Special features in the Bavarian Ministries

Recent Bavarian governments, including the present one, would say that the Greens are superfluous in Bavaria, since environmental issues, it is claimed have always been a high priority. Indeed the CSU does not accept that the Green Party MP Joschka Fischer was the first Minister for the Environment (*Umweltminister*) in Germany, when he was appointed in Hesse in 1982.

CSU politicians will remind you that their state government first introduced "a ministry for the environment," in its wider sense, in 1970: *das Staatsministerium für Landesentwicklung und Umweltfragen*, which has been a government department ever since.

It was then the first ministry of its kind in Europe, and the first incumbent was Max Streibl. Originally a large part of his task was concerned with planning and co-ordinating Bavaria's territorial infrastructure and nature protection in rural areas. More recently, of course, it has been involved in more standard environmental issues such as vehicle emissions and nuclear energy, especially after the disaster in Chernoby in 1986, which affected Bavaria more than any other (West) German Land.

Another new ministry is the one for Federal and European Affairs, created in 1988, reflecting the growing importance for Bavaria of the European Community, as it was, and the Single Market. This ministry is represented in Munich, Bonn, Berlin and Brussels. The international dimension of this Bavarian ministry means that there is also a valuable "spin-off" effect for the Bavarian economy.

Under Strauß the Bavarian state ministries of education (*Unterricht und Kultus*) and science (*Wissenschaft und Kunst*) were split, so there were ten ministries. When Streibl took over in October 1988, he combined them, but appointed two secretaries of state. Stoiber retained the same system, from June 1993 onwards, so that there are at present nine government departments (*Ressorts*).

There are three principles which the Land government must follow.

Firstly, the leadership principle (*Führungsprinzip*) means that the minister president himself may take decisions, for which he bears responsibilty vis-a-vis parliament, according to art. 47.2 BV. He lays down policy guidelines (*Richtlinien der Politik*).

Secondly, the collective principle (*Kollegialprinzip*) means that the government may take a collective decision. In the case of a tied vote, the minister president has the casting vote (art. 54 BV) and he must take responsibility for these decisions too.

Thirdly, the departmental principle (*Ressortprinzip*) means that a particular minister may take a decision for his or her area of competence and must accept the responsibility for it (art.51.1 BV).

The legislature

Bavaria once again stands out as an exception, in that it is the only federal state in the FRG which has a bicameral legislature. This particular political tradition can be traced back to 1818, when the Bavarian parliament consisted of two chambers, until 1918. In 1919 the Bamberg constitution returned to a unicameral system. In 1946 lengthy debates ensued and finally it was decided that Bavaria would have a second chamber: the Senate (*Senat*). Some observers claim that it is not really a second chamber in the full sense (see below).

The Landtag

The *Landtag* is the main legislative body. The Bavarian parliament is housed in the Maximilianeum, a magnificent edifice built in 1852 by King Maximilian II, who sponsored a foundation to support Bavaria's brightest grammar school achievers. We are reminded of the ubiquitous presence of the Catholic Church in Bavaria by the picture on the impressive facade of the building, high up in the centre, just below the statue of the goddess of peace. It depicts the founding of the Benedictine monastery at Ettal by Ludwig the Bavarian.

The *Landtag* is the main legislative body, elected every four years. It formally elects the minister president, and also elects its own presidium and the most important organ of the Bavarian judiciary, the Constitutional Court. Bavarian state electoral law determines that there are 204 deputies elected to parliament: 104 are elected "directly" via their constituencies and the remaining 100 come from the party lists. Art. 30, 31 BV state that all members of the Bavarian parliament have free use of all forms of public transport in Bavaria.

At present there are twelve parliamentary committees with between 17 and 24 members each. A Bavarian speciality is that the committees, with very few exceptions, meet publicly. School classes, the general public and the press regularly attend, following the principle of "transparent democracy" for the citizen, laid down by one of the fathers of the Bavarian constitution, Prof. Hans Nawiasky, who claimed: the people should to a certain extent participate in procedures and be able to keep a check on their representatives.[9]

The new *Landtag* set up in 1946 was built on foundations whose origins stretch back to the 13th century. The last of the old *Landtage* was in 1255[10]. Although Bavaria received its first attempt at a constitution in 1808 – the first step towards restricting the rights of a ruler by means of a written constitution – no proper parliament materialised until 1818. This body was based on the English and French models with a bicameral assembly, consisting of coucillors from the Empire (*Reichsräte*) and a chamber of deputies (*Abgeordnete*).

The former were members of the royal household (aged 18 or over), nobles, holders of inherited positions and other citizens over 21 appointed by the king. There were about 100 of these councillors. The deputies could be elected by citizens over 21 who had to fulfil residence, tax and property qualifications in order to gain the right to vote. They usually fell into one of five categories: nobled property-owners, clergy, university professors,[11] inhabitants of the cities and towns[12] and landowners. The number of deputies in parliament, who had to be at least thirty years old to stand, varied between 115 in 1819 and 163 in 1918.

The present-day parliament

Modern parliamentarians in Bavaria enjoy indemnity (art. 27 BV) and the

---

[9] op.cit.Roth, p.82.

[10] Der Bayerische Landtag, Sonderteil der Verfassung des Freistaates Bayern, p.213, published by the Bayerische Landeszentrale für politische Bildungsarbeit. 1993.

[11] Some of them came from Landshut University, which was moved to Munich in 1825. See chapter 2.

[12] The three largest cities in present-day Bavaria, namely Munich, Nuremberg and Augsburg, voted, even in those days, for deputies directly, since they were in a privileged position for political and historical reasons. Reinhold L. Bocklet (ed.) Das Regierungssystem des Freistaates Bayern. Band II. Munich. Vögel. 1979, p.32.

protection of immunity (art.28 BV) and sit in groups according to parties (*Fraktionen*). A party had to have a mininmum of ten members in order to qualify for the status of a *Fraktion*, or parliamentary party. The figure of ten was recently reduced to five. This was a vital change because *Fraktionsstatus* guarentees secretatial assistance and support services, participation in the presidium and in the committees, including the right to chair a committee, as well as speaking time.

Small parliamentary parties, such as the Bavarian FDP, found it extremely difficult when they entered parliament but did not receive the status of being a *Fraktion*. Parliamentary life was still problematical, even when they did have that status, because only twelve or thirteen deputies had to shoulder the burden of sitting on several committees etc[13]. On four occasions since 1946 the Bavarian Liberals gained less than ten seats in parliament, and on three other occasions they failed to enter parliament. More recently, the Greens have faced a similarly difficult situation. After the 1986 state elections the Bavarian Greens gained fifteen seats in parliament in Munich and twelve during the following period of legislature (1990-94).

The main function of the Bavarian *Landtag* is of course passing legislation. In Bavaria it is possible for the people themselves to act as legislators, via the device of the refendum (see chapter four.) The state budget law is the only one where no referendum is permitted.

Laws are normally passed by parliament (art. 72.1 BV). Most laws are proposed by the state government, but may also come from the Senate (see below) or the floor of the house. A special feature of the legislature in Bavaria is that draft laws must go to the second chamber for consideration. The government then decides if it wants to accept or reject any objections the Senate may have. Normally a bill is given two readings – a third may be requested – and the minister president must sign the law before it goes into the statute book.

Art. 31 GG, whereby federal law overrules *Land* law, means of course that certain restrictions are placed on the Bavarian legislative process. It must conform to both the German and Bavarian constitutions and follow strict procedures.

---

[13] This point was explained to the author during an interview on 15.6.1982 with Hans-Jürgen Jäger, then leader of the Bavarian FDP parliamentary party in his offices in the Maximilianeum in Munich.

The Senate – a unique feature

The Senate (*der Senat*) is one of several unique features in the Bavarian polity. Bavaria is the only federal state in the FRG which has a second chamber in its legislature. Some commentators debate whether the Bavarian Senate really can be considered a fully fledged second chamber, since it is not comparable with the Bundesrat, for instance[14], and since it has *no* right of veto on legislation passed by the *Landtag*.

As we have already seen, Bavaria has a long tradition of bicameral parliaments. In the lengthy debates in the immediate post-war period, the CSU and Hans Nawiasky were in favour of a second chamber. The SPD was against the idea and the FDP was unenthusiastic. After considerable debate it was decided in 1946 to establish a *Senat* (literally: a council of elders) as one of the four highest organs of state, along with the Landtag, the State Government and the Constitutional Court (art.64 BV).

It is possible that the American Occupation Forces influenced the decision to set up a second chamber, as Karl Bosl suggests, given their familiarity with a federal and bicameral system[15]. It is however more likely that, as usual, Bavarian traditional thinking and a strong desire to forge links with past political systems was the more influential factor.

On 26 August 1946, during discussions at a plenary session of the Constitutional Assembly in Munich, Dr. Hans Ehard, who was Bavarian minister president from Dec. 1946 until Dec. 1954, said that the Senate was above all intended as a stabilising factor. It was envisaged as a counter-balance to the purely party-political thinking and way of representing the people provided by the *Landtag*. The Senate is supposed to guarentee protection against hurried or ill-conceived decisions. It was seen by some as the "moral conscience" of the *Landtag* and State Government.[16]

---

[14] "Der bayerische Senat ist nicht mit dem Bundesrat zu vergleichen." Hans-Helmut Rösler, Landesgeschäftsführer der bayerischen F.D.P. discussed the role of the Senate with the author at the party's headquarters in Munich on 20.10.1993.

[15] Karl Bosl's contribution "Neubeginn – die Gründung der zweiten bayerischen Republik 1945/46" in Bavaria Felix, ed. by Bernd Rill. Schulz. Percha. 1986. p.16.

[16] "Der Senat als moralisches Gewissen." Rainer Roth. Freistaat Bayern. Die politische Wirklichkeit eines Landes der BRD. Bayerische Landeszentrale für politische Bildungsarbeit. München. 1986, p.271.

According to the constitution – various points are covered in art.34-42 BV – the Senate is a body representing social, economic, cultural and local interests in Bavaria, which consists of 60 members, who must be at least 40 years old. They are elected according to democratic principles from within the areas they represent. The only exception to this are the five representatives from the religious communities who are appoinnted (art.36.1 BV).

The sixty representatives are selected in the following way:

11 from agriculture and forestry
11 from trade unions
 6 from communes and local interest groups
 5 from industry and commerce
 5 from the craft trades
 5 from cooperatives
 5 from the religious orders
 5 from welfare organisations
 4 from the free professions
 3 from universities and academies

The senators do not vote as a group or a *Fraktion*. Each individual senator makes his or her own decision. This is felt to be a strength. For the discussions on draft laws, however, specialist committeees exist. Alongside the main committee, there are six committees, dealing with elections, finance and budget concerns, legal and constitutional matters, cultural policy, economic policy and the committee for social, health and family policy.[17]

These senators remain in office for six years. This means that there is an overlap with the *Landtag*, which is elected every four years. One third of the 60 Senate members are replaced every two years. In this way the Senate is a permanent body providing an element of continuity.

It has the right to propose draft laws and bring initiatives straight to the *Landtag* and these must be laid before parliament. The Bavarian government is supposed to ask for the opinion of the Senate on "all important matters" (art.40 BV). It **must** do so in the case of the state budget law, changes in the constitution and laws which are put before the people for a decision.

Any laws passed by the Bavarian parliament are presented to the

---

[17] See information brochure "Bayerns politisches System" in the series "Gesellschaft und Staat", produced by the Bayerische Landeszentrale für politische Bildungsarbeit 4/89, p.6.

Senate before publication. The Senate has the right to raise objections accompanied by reasons within one month, unless the matter is considered urgent, in which case the period is reduced to one week. *The Landtag decides whether it wishes to take the views expressed by the Senate into consideration or not* (art.41 BV).

The role and function of the Senate in the Bavarian political system may be summarised as follows. Firstly, the Senate should react to and advise upon draft laws, especially the state budget, drawing on its wide experience from various walks of life outside politics. Secondly, it may from time to time propose draft laws (the right of initiative) to the *Landtag*.

Thirdly, it should advise on all changes in the constitution and any bills put to the Bavarian people as referenda. Fourthly, it checks all laws passed by the Landtag and makes any comments or objections it feels appropriate. It has *no* veto rights on legislation – only the right to advise, comment and raise objections. The Senate can recommend but not compel the Landtag or the state government to act upon this advice.

Members of the Senate do not receive any remuneration for their work, only expenses. In order to be quorate for decisions, 31 of the 60 members must be present. Regardless of any theoretical debates on whether the Senate is a genuine second chamber or not, it *is* a legislative body and a part of the Bavarian polity as one of the highest organs of state (art.64 BV). Although critics say that referring bills to the Senate wastes time, experience shows that the senators have often made valuable contributions, based on their practical, non-political stance.

Between 1948 and 1990 the Senate gave expert advice on 435 draft laws put forward by the government and almost 300 draft laws proposed from the floor of the *Landtag*. That amounted to over half of the 1,228 laws passed. Of the small number of draft laws where the Senate raised an objection (12 per cent), the Landtag amended its proposals in about half the cases, even though it did not have to do so. Over the same period the Senate made use of its right to initiate draft laws 37 times and on 885 occasions it expressed an opinion on the constitutionality of laws and/or lodged a complaint (art.64 BV)[18].

---

[18] All figures taken from op.cit. Roth, p.92/93.

The Judiciary

The third pillar of the Bavarian political system, the judiciary, also owes its present structure to traditions of the past. The most important court is the Bavarian Constitutional Court (*bayerischer Verfassungsgerichtshof*). The constitution deals with this aspect of jurisdiction in art.60-69 BV. This court is situated in Munich and is the *Land* equivalent of the Federal Constitutional Court in Karlsruhe.

In the middle of the 19th century, under a constitutional monarchy, and later in the Weimar Republic, Bavaria possessed a similar type of special court known as the *Staatsgerichtshof*, which was first established by a law passed on 30 March 1850. Its sole task was to deal with complaints concerning ministers. In 1919 the Free State of Bavaria set up another such court, which took on the same task, as well as dealing with constitutional matters. The present Constitutional Court in Bavaria, constituted in 1947, is a completely independent body operating exclusively on a constitutional level.

Every inhabitant of Bavaria, and that includes foreigners because they have a relationship to the territory of the Bavarian state by virtue of residence, has the right to call upon the protection of the Bavarian Constitutional Court, if they feel their rights have been violated either by a court or by the authorities.

A special feature of Bavarian constitutional law is the so-called right of "popular complaint" (*Popularklage*). According to art.98 BV, the Constitutional Court has the right to declare null and void any laws or decrees which unconstitutionally restrict a basic right.

The name derives from the fact that anyone may appeal to the Court in this way (art.53.1 of the law on the Bavarian Constitutional Court). As a matter of principle, this procedure does not entail any costs. It is esimated that around fifteen per cent of the "popular complaints" made are successful.[19]

The separation of the judiciary and state administration in Bavaria dates back to 1861. The law courts (ordinary jurisdiction) are divided into three strands: civil, criminal and voluntary jurisdiction.

---

[19] See Horst Säcker. Der Bayerische Verfassungshof, a contribution in Reinhold L. Bocklet (ed.). Das Regierungssystem des Freistaates Bayern. Band II. Bayerische Landeszentrale für politische Bildungsarbeit. Munich. 1979. p.451.

There are also four main categories of jurisdiction: administrative, fiscal, labour and social insurance. Bavaria has 22 state or regional courts (*Landgerichte*) and, at the lowest level, 72 local courts (*Amtsgerichte*). Another special feature of the Bavarian system is the existence of a Higher Regional Court (*Bayerisches Oberstes Landesgericht*), whose history stretches back to the 15th century.

The history of the Bavarian party system

Several acknowledged authorities writing on Bavarian historical and political traditions, for example Mintzel, Gebhardt, Falter, Amery, Bosl, all agree that it is only in recent decades that a single political culture has emerged in Bavaria. As has been shown already, when the modern state of Bavaria was formed under Montgelas in the early 19th century, it brought together a mixture of many political cultures. The three broad traditional zones of Old Bavaria, Franconia and Swabia were not only very different from each other, they also each contained varying political and historical cultures of their own.

From around 1819 onwards one or two loose cultural associations began to form in Bavaria, bringing together like-minded compatriots. From around 1831 onwards a few ideologically inspired amalgamations occurred, but it was not until 1849 that more identifiable loose political groupings began to appear. This new phenomenon was influenced by the first (unsuccessful) attempt at the *Paulskirche* in Frankfurt in 1848 to unite the German states.

Bavaria was represented in Frankfurt by 71 mainly moderate liberal deputies, who did not sit as one separate group in the German National Assembly in St Paul's church, but instead joined the political groups which were at this time just beginning to form.

On 4 June 1848 a new electoral law was passed in Bavaria and a *Landtag* was elected with four main political groups emerging: a group of right-of-centre Conservatives, two Liberal groups and a radical left-wing group. One of Germany's first political parties was the German Progressive Party (*Deutsche Fortschrittspartei*), formed in 1861. This was followed two years later, on 15 March 1863, by the formation of an equivalent party in Bavaria. The party was founded by Karl Brater, liberal deputies Marquardt Barth and Josef Völk and a lawyer from Erlangen called Heinrich Marquardsen.[20]

The Bavarian *Landtag* of 1869 was the first one which contained clearly identifiable political parties as such. It included the Bavarian Patriots' Party (*Bayerische Patrioten*), which was the first Christian (conservative) party to exist in Bavaria. There was, however, a separate group of Conservatives, who were in fact the largest group in Bavaria in the 1870s and the 1880s. The early Bavarian party system was basically a Christian-conservative-liberal one, with the strength of political liberalism, both in Germany and Bavaria being considerably weakened by the split which occurred so soon after its establishment.

The Bavarian Patriots were the start of a movement representing political Catholicism in Bavaria, continued by the Catholic Centre Party (*Zentrum*) and later in the Weimar Republic the Bavarian People's Party (*die Bayerische Volkspartei*). This tradition was re-established after 1945 by the formation of the CSU (*die Christlich-Soziale Union*).

---

[20] Details taken from Friedrich Henning's contribution "Entwicklungen des politischn Liberalismus in Bayern," Nachwort zur Dissertation von Berthold Mauch: Die bayerische FDP. Porträt einer Landespartei 1945-49. Olzog. Munich. 1981. p.102.

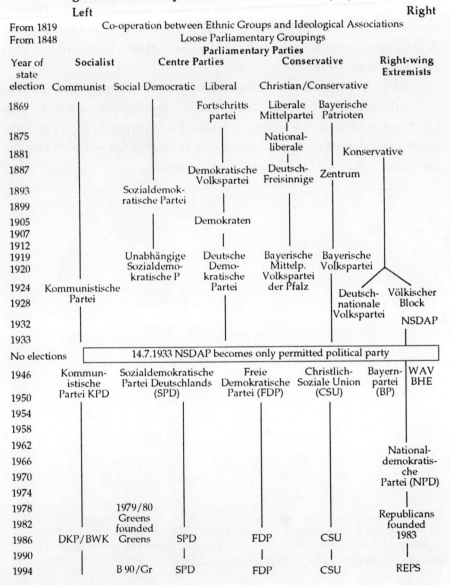

Fig 3.2 The Development of the Bavarian Party System

4

A new Constitution for a "new" State

Territorial continuity

Following the total collapse on 8th May 1945, the American Occupation Forces were put in charge of Bavaria, which in terms of territory remained the same as it had been when it was disbanded under Hitler's *Gleichschaltung* of 1933, except for the loss of the Palatinate and, initially, the tiny area of Lindau.

In August 1945 one of the heads of the department for government structure in OMGUS (Office of Military Government of the United States), James William Kerr, made a tour of Germany. Afterwards he reported that Bavaria was the only region that could continue on the basis of previous governmental structures in an effective manner, and whose territory and administration were integrated.[1]

Heinz Laufer, writing on German federalism, comments that Bavaria possesses its own distinctive political culture and well established traditions, and that, with the possible exception of Hamburg and Bremen, nowhere are the emotional ties and pronounced feelings of identity as strong as they are in Bavaria.[2]

---

[1] Walter Siegel. Bayerns Staatswerdung und Verfassungsentstehung 1945/46. Bayerische Verlagsanstalt. Bamberg. 1978. p.24.

[2] Heinz Laufer. Das föderative System der BRD. Bayerische Landeszentrale für politische Bildungsarbeit. Munich. 1981. p.178.

It is crucial to realise that, despite the many problems in the immediate post-war situation, Bavaria was able to re-establish itself much more easily than most other regions. Even in 1946 Bavaria could look back on 140 years of existence as a modern nation state. Modern Bavaria (*Neubayern*) was developed in the Napoleonic era in the early nineteenth century, not to mention Bavaria's thousand-year history.

Re-instatement of democracy

The American Military Governor of Bavaria, on 25th May 1945, entrusted the task of setting up a provisional Bavarian government to Fritz Schäffer, a former member of the Bavarian People's Party (BVP), who later became a founder member of the CSU. The Americans were, however, dissatisfied with Schäffer's efforts, especially regarding de-nazification, and at the end of September they replaced Schäffer with Wilhelm Hoegner from the SPD.

Between 8 March and 24 June 1946 a small committee led by Hoegner was given the task, by the Americans, of collecting all documents which might be of use in preparing a draft for a Bavarian constitution. It should not be forgotten that at that time neither the *Reich* nor the *Bund* existed. The main objectives were to try to avoid the errors which had led to the downfall of democracy under the Weimar constitution.

In preparation for democratic elections, the U.S. Occupation Forces began to license political parties from January 1946 onwards, first at local, then at state level. On 30th June 1946 elections for a constitutional assembly were held, based on the principles of proportional representation in a universal, equal, direct and secret vote. The conservative Christian Social Union (CSU)[3] gained 58.8 per cent (109 seats), and the Social Democratic Party (SPD) 28.8 per cent (51 seats). The Economic Reconstruction Party (WAV), which existed only in Bavaria, and the Communist Party (KPD) each gained just over five per cent (eight and nine seats respectively); the liberal Free Democratic Party (FDP) polled only 2.5 per cent (three seats). There were 180 seats in the first post-war Bavarian assembly.

The assembly appointed a committee consisting of twenty-one members to prepare a draft document and discussions on a new Bavarian con-

---

[3] More details on the political parties established in post-war Bavaria are given in Chapter 5.

stitution began. Hans Nawiasky, a professor of constitutional law, had been involved in drawing up the Bavarian constitution in 1919. He and Wilhelm Hoegner, both of whom sought exile from the Third Reich in Switzerland, were two of the most important "fathers" of the new Bavarian constitution. Their time in Switzerland influenced them with respect to the use of the referendum, which is examined later in this chapter.

On 26 October 1946 the constitutional assembly, at a plenary session, meeting in the unheated main lecture theatre of Munich university, voted in favour of accepting the new constitution by 136 votes to 14.[4] General Clay, the American military governor of Germany, had accepted the draft constitution two days earlier, with the proviso that Bavarian citizenship (referred to below) also had to mean German nationality.

Deep-rooted constitutional traditions

As was seen in the previous chapter, Bavarian constitutional traditions stretch back a very long way. The first known book of Bavarian law, the Lex Baiuvariorum, was completed in the seventh century, around the year 630, after several decades' gestation. Several other important documents followed (e.g. the Schnaitbacher Urkunde in 1302 and the Ottonische Handfeste in 1311).

Bavaria's first real constitution or *Verfassung* came in 1818, ten years after Montgelas's preparatory *Konstitution*, which set some basic ground rules. The 1818 Bavarian constitution is generally recognised as the first fully successful attempt to establish the foundations of a modern parliamentary democracy in Bavaria. It is interesting to note that just one year before (5th June 1817) a Concordat between the Bavarian state and the Roman Catholic Church was signed. The Bamberg Constitution of 1919 for the Free State of Bavaria was subordinated to Germany's Weimar Constitution.

It must be remembered that, after the Second World War, the new Bavarian constitution came into force (December 1946) nearly two and a half years before the one for the Federal Republic (May 1949). This being so, Bavaria was determined to anchor its autonomy (*Eigenstaatlichkeit*) as firmly as possible, in anticipation of a new German state.

---

[4] Figures taken from Helmut Hoffmann. Handbuch zur staatspolitischen Landeskunde der Gegenwart. Olzog. Munich. 1985, p.47.

A second Free State of Bavaria

The Bavarian constitution, which was approved by the Bavarian people in a referendum on 1st December 1946, makes reference in its preamble to Bavaria's thousand-year history. Bavaria was, in 1946, a fully grown state with its own finely tuned sense of identity.

The very first article of the constitution states that Bavaria is a Free State. The German term, *Freistaat*, was first used in 1919 (see chapter two). Wilhelm Hoegner explained in 1946 that this term was simply a translation of the word *Republik*, which was still felt by some to be a foreign word. This is referred to in a note in the Bavarian constitution, which later also added that the re-introduction of a monarchy would contravene article 28(1) of the Basic Law – the Grundgesetz or provisional constitution of the Federal Republic, signed on 23rd May 1949.

Even in 1919 Bavaria was not the only region to be called a Free State, and in the new Germany established in October 1990 the *Land* of Saxony is known as *der Freistaat Sachsen*. The author has observed on several occasions that, particularly CSU politicians speaking in the Bavarian parliament in Munich or the Bundesrat in Bonn, make frequent use of the term *der Freistaat Bayern*, as if wishing to emphasise Bavaria's special status by using its "full title."

Nevertheless the term Free State of Bavaria has no special status vis-a-vis other German *Länder*, regardless of the psychological importance some Bavarians may attach to it. It is, strictly speaking, no different from the terms used by the Free Hanseatic city states of Hamburg and Bremen. That is, however, not the way many Bavarians see it, especially the inhabitants of rural *Altbayern*.

Bavarian citizenship

As part of the desire to emphasise Bavarian autonomy, the constitution of 1946, in art.6, spoke of "Bavarian citizenship" (*bayerische Staatsangehörigkeit*). This was first commented upon by the US Occupation Forces (see above). It is, though a minor point, to some extent a typical illustration of how Bavaria has often wanted to be different. It may be of psychlogical significance to some Bavarians, but it exists in theory only, since the Basic Law, when it was promulgated in 1949, stated in art. 31 that federal law overrules *Land* law. Art.74 (no.8) GG and art.16 and 116 GG also deal with the question of citizenship. All Germans in the Federal Republic possess German nationality, and there is officially no such thing as nationality or citizenship of the individual *Länder*.

There can, however, be little doubt that some of the older members of the Bavarian population agreed, and may still agree, with those who included the idea of Bavarian citizenship in the 1946 constitution, known colloquially as the "Bavarian passport" (*Bayernpaß*). Given Bavaria's long tradition as a separate nation state and kingdom, it is no wonder many Bavarians felt they were first and foremost Bavarians, and Germans second. This was especially true in 1946 when the constitution was drawn up, before the Federal Republic was founded. As will be discussed later, it is arguably still true, even today, for a small portion of the Bavarian people.

Bavaria introduces "direct democracy"

The new constitution envisaged an element of representative or direct democracy. It stipulates that the Bavarian people have the right to request a referendum (see below). In addition, any change in the Bavarian constitution requires the approval of the Bavarian people. This stipulation, which is firmly anchored in the Bavarian constitution – art.75.2 BV (*bayerische Verfassung*) – contrasts with the Basic Law, which according to art.79 GG (*Grundgesetz*) requires only a two thirds majority of the Upper and Lower houses of parliament in order to amend the German constitution.

There have been only five amendments to the Bavarian constitution since 1946. On 22 July 1968 a new formulation of art.135 BV, concerning the introduction of the Christian non-denominational school, was approved. The second amendment concerned the reduction in the voting age – from 21 to 18 – on 15 June 1970 (art.7.1 BV). The age at which you may stand for election was reduced, from 25 to 21 (art.14.2 BV), at the same time. On 19 July 1973 Bavarian electoral law was further amended. This time the change dealt with constituencies and also changing Bavaria's ten per cent clause to the standard five per cent clause. On the same day a fourth amendment was introduced: freedom of broadcasting. That was the only example so far of a successful change in the constitution which resulted exclusively from a direct initiative from the Bavarian people. The fifth change came on 20 June 1984. It was in connection with the protection of nature and the environment.[5]

Use of the referendum

The fact that the Bavarian constitution promotes maximum representation of its citizens via "direct democracy" by stipulating the availability of the referendum was mainly due to the experiences of Hoegner and Nawiasky whist in exile in Switzerland. This aspect of the new democratic system has secured two things. Firstly, the people of Bavaria have always been directly involved in any constitutional amendments. Secondly, far fewer changes have been made to the Bavarian constitution than to the German constitution (only five compared to 38 changes to the Basic Law so far).

The following procedure is laid down: a request for a referendum (*Volksbegehren*) may be made on any issue – with the sole exception of the Bavarian state budget (art.73 BV). The Bavarian constitution grants the Bavarian people the right to be an organ of the state by participating directly in the democratic process. They may avail themselves of this direct influence via the device of the referendum or plebiscite (*Volksentscheid*) in three ways.

Firstly, the people may initiate a new law by a request for a referendum (art.74.1 BV) and, if successful, then participate in a referendum (art.72 .1BV). Secondly, the people have the right to hold a referendum on any proposed change in the constitution, which must first have been passed by a two thirds majority in parliament (art.75.2 BV). Thirdly, the people have the right, according to art.18.3 BV, to dissolve parliament if one million enfranchised citizens demand it.

The initial application for a request for a referendum must be directed to the Bavarian Ministry of the Interior. It has to be accompanied by supporting reasons and signed by 25,000 enfranchised citizens. The proposed bill or constitutional amendment is then scrutinised by the ministry and the Bavarian Constitutional Court. If it is deemed acceptable, ten per cent of

---

[5] These amendments are covered in detail in the latest edition of the Bavarian constitution, available from the Bayerische Landeszentrale für politische Bildungsarbeit in Munich.

those entitled to vote at the last Bavarian state elections must support the request with their signatures.

It then goes forward to both houses of parliament and the Bavarian government and may also be referred again to the Constitutional Court. The next step is for a referendum to be held. At the referendum itself the people vote by answering only "yes" or "no" to the question. The majority of votes cast must be "yes" votes. If there is an equal number of "yes" and "no" votes, the request is rejected.

Not everyone of course is in favour of referenda, which are thought by some to be the weapon of the demagogue and the dictator. In a modern democracy the device of the referendum should be used sparingly. In the case of Bavaria it has been. There have been only eleven requests for referenda since the war, four of which failed.

In addition to the five changes to the constitution already mentioned, there was the original referendum on 1 December 1946, in which 75.7 per cent of the electorate turned out to vote on the new Bavarian constitution; it was accepted with a vote of 70.6 per cent in favour.[6] In 1967 an unsuccessful attempt to abolish denominational schooling was made – only 9.3 per cent voted in favour of a referendum, so no referendum could be held. (The 1968 one was successful and resulted in a change in the constitution.)

In 1971 only 3.7 per cent supported a request regarding territorial reform. Six years later there was an unsuccessful request for free books and materials at school. In the same year, 1977, a request for a referendum on a change in the representation in the second chamber of the Bavarian parliament also failed to attract the necessary ten per cent support.

Finally, the most recent example started with a suggestion in November 1988 for a change in the law, when a citizens' initiative group drafted a proposal on a waste disposal scheme. The request for a referendum did not come until June 1990, when it was supported by 12.8 per cent of those entitled to vote. At the actual referendum, however, on 17 Feb. 1991, it was a choice between this proposal and a government bill, which had materialised in the meantime. The government bill received 51 per cent support and therefore was adopted, whilst only 43.5 per cent supported the

---

[6] There were 2,090,444 votes for and 870,135 against. op. cit. Hoffmann, p.48.

original request. This was the first time that the Bavarian people had been active as "legislators" via a referendum.[7]

Idiosyncratic features of the constitution

The third section of the Bavarian constitution (art.34-42) deals with a unique feature of the Bavarian political system, the Senate or second chamber of parliament. No other federal German state has a second chamber. Bavaria's *Senat* is, however, not comparable with the *Bundesrat*, the German Upper House in Bonn.

Unlike the German constitution, there is no constructive vote of no confidence in Bavaria. Article 44.3 BV does make provision for the possible resignation of the minister president, if confidence has been lost in him.

The Bavarian constitution uses the terms *state* government, *state* ministers and *state* secretaries, as it sees itself as more than simply another *Land*. Unlike other federal states in Germany, which usually refer to the *Landesregierung* etc., Bavaria again is the exception to the rule by referring to the *Staatsregierung*, etc. Even today these terms can be heard regularly in radio and television broadcasts in Bavaria. Yet another little reminder that Bavaria thinks of itself as a (separate) state with a special position – the Free State of Bavaria.

Bavaria is also the only *Land* where not only the Minister but also his or her second-in-command, the *Staatssekretär*, sits in the cabinet as a member of the government (art.43). The Bavarian government and parliament are subject to the authority of the Bavarian Constitutional Court (art.61 BV). This is Bavaria's highest court of law for all constitutional matters and plays a crucial role in the Bavarian judiciary.

In 1946 Bavaria also had a unique electoral system, covered in art.14 of the constitution and in more detail in the electoral law. It was unique because Bavaria was, up until 1973, the only state in the Federal Republic of Germany which had a ten per cent clause for elections to its state parliament (*Landtag*). It also used to be the only Land where the voter had two votes in a state election (*Landtagswahl*). The electoral system used in local elections in Bavaria also possesses some special features, discussed below.

---

[7] This most recent case is dicussed in Rainer A.Roth. Freistaat Bayern. Politische Landeskunde. Bayerische Landeskunde für politische Bildungsarbeit. Munich.1992, p.69.

Art.160 of the Bavarian constitution concerns nationalisation of property. This was a somewhat unusual inclusion, given that the CSU, which had an absolute majority in the constitutional assembly, was normally opposed to such policies. It was, however, included as a "sop" to the SPD, the second largest party, in order to secure their agreement to art.135 BV, which stated that Bavaria's schools would be segregated according to denomination (Protestant or Catholic). As will be seen later, art.135 proved extremely controversial and was eventually amended in 1968, after many furious debates in parliament. On the other hand, no use has ever been made of art.160.

In Bavaria the state government, according to art.43 BV, consists of the minister president, the state ministers and the state secretaries. They are responsible, under art.43, 47 and 51 to the Bavarian parliament. Great importance was attached to limiting the power invested in the the people's representatives by means of regular elections and, where requested, referenda.

The Bavarian coat of arms

The constitution refers, in art.1, to the official state colours of white and blue and the Bavarian coat of arms. The shield, held by two lions, is divided into quarters. On the top left is a golden lion on a black background, which was originally the symbol of the Palatinate Counts on the Rhine. For centuries this was the common emblem of the Old Bavarian and Palatinate House of Wittelsbach. Nowadays the Palatinate lion serves as a reminder of the Bavarian district of the Upper Palatinate.

The top right-hand quarter of the shield bears the Franconian "Rake" in the colours of Franconia, red and white. The bottom left quarter displays a blue panther on a white background. This was originally in the coat of arms of the Palatinate Counts of Ortenburg and later the Wittelsbach family. It now represents the districts of Lower and Upper Bavaria. The bottom right quarter, three black lions one above the other on a yellow background, is taken from the old coat of arms of the House of Hohenstaufen, once Dukes of Swabia. In this way that part of present-day Swabia which belongs to Bavaria is represented. The white and blue central shield in the middle of the coat of arms was once used by the Counts of Bogen, later adopted by the House of Wittelsbach. Today it represents Bavaria as a whole – as the "small state coat of arms" (*Raute*).

The Bavarian anthem

Bavaria also has its own "national" anthem or hymn (*Bayernhymne*), which again appears to be of some psychological significance to part of the Bavarian population. The author once witnessed how the right-wing Republican Party finished a Bavarian election rally by singing the German national anthem, including the first two, now "forbidden" verses of the original version. On the following day the CSU, not to be outdone, at the close of their rally in front of the Munich town hall, brought along the ubiquitous Bavarian *Blaskapelle* and played the Bavarian hymn. They were the only party to do so.

The *Bayernhymne* dates back to the middle of the nineteenth century. It was written by Michael Oechsner (1816-93) and the music for it was composed by Max Kunz (1812-75). The original version consisted of three verses: the first verse is dedicated to the state of Bavaria[8], the second to the Bavarian people and the third to the Bavarian king. After the collapse of the monarchy in 1918, the third verse became obsolete.

On 18 July 1980 the Bavarian minister president, F.J. Strauß, decreed that the original version of the first two verses, first approved by the Bavarian cabinet in 1953, should be used. The Bavarian hymn is still heard on Bavarian radio and was taught and sung in schools from about 1953 onwards[9]. This is perhaps just another of the little details which, whilst being of little or no significance to outsiders, mean a lot to some Bavarians

---

[8] Gott mit dir, du Land der Bayern, deustche Erde, Vaterland. It went on to praise the white and blue colours of the Bavarian flag. From around 1952/53 onwards the Bavarian hymn was taught and sung in Bavarian schools. The words were altered slightly by Josef Maria Lutz, in order to update it.

[9] Taken from an appendix to a recent version of the Bavarian constitution (1993), p.95/96, published by the Bayerische Landeszentrale für politische Bildungsarbeit.

and, when added to a number of other factors, enhance Bavaria's claim to special status.

Another such matter is the Bavarian medal or Order of Merit (*der Bayerische Verdienstorden*), in the shape of a Maltese cross. Intoduced in 1957, it was created in order to honour and recognise outstanding achievements or service to the Free State of Bavaria or the Bavarian people. A long Bavarian tradition was prolonged in 1980 by the creation of an award for services to art and science (*Maximiliansorden*).[10]

Professor Nawiasky insisted that democracy had to be "restricted rule." He described the determining principles of the Bavarian constitution, which helped to establish political order in the immediate post-war period, as moderation and continuity. The new foundations of a democratic state were to represent the exact opposite of the horrific excesses of the failed National Socialist experience.

At the same time Bavaria, in 1946, did everything possible to emphasise its independence and special position, based on its long and enduring traditions, its thousand-year history, territorial continuity and strong emotional ties.

These had, over many centuries, produced a feeling of being a close-knit, self-contained, independent unit with a highly developed sense of being a nation state. The new constitution in 1946 tried to build on the past – a past of which Bavaria was extremely proud – laying the foundations for a "new" German *Land*, the Free State of Bavaria, which, given its history, was by no means entirely new.

The Bavarian electoral system

On the one hand the electoral system operated in Bavaria could be considered not entirely fair, in the strictest sense of the term, because votes are counted and seats are allocated on the basis of the seven districts, not the whole of Bavaria. On the other hand the system does make a genuine attempt at offering the voters as much democratic choice as possible in selecting the actual candidates who represent them. It is yet another unique aspect of the Bavarian political system.

---

[10] Bayerische Maximiliansorden für Wissenschaft und Kunst. op.cit. Hoffmann, p.58/59

The 1946 Bavarian constitution states in art.2(1) that the power of state eminates from the people. This invokes the principle of the people's sovereignty; the second paragraph of the article states that the people express their will via elections.

Voting rights

Art.14(1) BV states that the election of deputies is universal, equal, direct and secret. The West German Basic Law (*Grundgesetz*) in 1949, in art.38, used these same four terms, plus the term free, which is included automatically in the Bavarian stipulations by virtue of the others.

Elections in Bavaria take place according to an **adjusted** system of proportional representation (see below). Active voting rights are granted to any German over 18 who has been living in Bavaria for at least six months. It is possible for this requirement to be lengthened to one year, according to art.7(3) BV.

Passive voting rights for federal elections are granted to anyone over 18, but in order to be eligible to stand for election at a Bavarian state election (*Landtagswahl*), or to stand as mayor or *Landrat*, 21 is the minimum age. For the local councils (*Gemeinderat and Kreistag*) anyone over 18 may stand for election. Art.44(2) BV stipulates that anyone standing for the office of either Bavarian minister president or a member of the Senate (art.36 (2) BV) must be at least 40 years of age.

Ten per cent clause

One of the unique features of the Bavarian electoral system introduced in 1946 was the ten per cent clause; every other West German *Land* has always had a five per cent clause. Several independent sources confirmed that one of the original reasons, though never explicitly stated, behind the introduction of a ten per cent hurdle in Bavaria was to exclude the Communists from parliament in Munich.

The ten per cent hurdle was applied *not* to the whole state of Bavaria but to any one of the seven districts. It therefore became crucial to small parties – for example the FDP – to muster support in one of the seven districts. Hence the reference to an "adjusted" system of PR.

In fact the Bavarian FDP always managed to clear the ten per cent hurdle at each of the state elections between 1946 and 1962, but only in one of the seven districts, namely in its stronghold of Central Franconia, gaining between 11 and 13.3 per cent.

In 1966, however the FDP vote fell to 9 per cent in Central Franconia at the Bavarian state election, as the new right-wing National Democratic Party (NPD) gained 12.2 per cent of the vote in that district, replacing the Liberals in the Munich parliament during the 1966-70 period. At the 1970 *Landtagswahl* the NPD disappeared again and the FDP returned to the Maximilianeum, the Bavarian parliament after polling 12.4 per cent in Central Franconia.[11]

On 19.7.1973 there was a change to the Bavarian ten per cent clause. Until then only parties which gained at least ten per cent of the total first and second votes in any one of the seven districts (*Regierungsbezirke*) had been able to enter parliament. It was amended to a five per cent clause (applied to the whole of Bavaria), which brought it into line with all other federal states. In fact the liberal party chairman, Thomas Dehler, had objected to the Bavarian ten per cent clause during some of the early discussions on the Bavarian constitution in 1946, but to no avail.[12]

The two-vote system

Another unique aspect of the electoral system enacted in post-war Bavaria was the two-vote system. Although the Bavarian electorate was given two votes each at *Land* elections, the system differed from that used for federal elections. Until more recently, every other federal state in Germany had a one-vote system, and this is still the case in most of the federal states, though not all nowadays. At the state election in Hesse in February 1995, for example, each voter had two votes.

In a Bavarian parliamentary election those entitled to vote cast their first vote for a directly elected constituency candidate, on a simple majority or first-past-the-post system (*relative Mehrheitswahlrecht*), as in a German

---

[11] Figures taken from various files in the archives of the Friedrich Naumann Stiftung, formerly in Bonn, now situated in Königswinter.

[12] Walter Siegel. Bayerns Staatswerdung und Verfassungsentstehung 1945/46. Bayerische Verlagsanstalt. Bamberg. 1978. p.139.

federal election or a British general election. In terms of actual terminology used, Bavaria is again an exception to the rule. Its constituencies are called *Stimmkreise* rather than *Wahlkreise*. In Bavaria each of the seven government districts becomes an electoral district, called a *Wahlkreis*, for the purposes of a Land election.

However the second vote is not cast simply for a party list, as in a federal election; the Bavarian voter has the opportunity of altering the order of the candidates presented on the party lists by placing a cross against the name of an individual candidate on one of the party lists, which are based on the seven electoral districts or *Wahlkreise* rather than the whole of Bavaria.

Many voters avail themselves of this opportunity, although a cross may simply be placed at the top of a party list, without indicating a preference for any particular candidate. An examination of what happens in practice reveals that the order of the candidates on party lists is frequently altered by the voters.

In fact the voters usually avail themselves of the opportunity of indicating an individual preference within a party list to such a great extent that some parties place their first ten candidates only at the top of the list. The remainder are placed simply in alphabetical order, since the party knows that the order of candidates is probably going to be altered anyway – often quite radically – by the wishes of the voters.

A particularly spectacular example of this occcurred in Bavaria in 1962 when the FDP candidate Dr Hildegard Hamm-Brücher was moved from seventeenth to first position on the party list, polling some 45,000 votes in all.

This aspect of the Bavarian system certainly goes a long way towards meeting the criticism sometimes levelled at the federal electoral system that the party hierarchy has too much power, because it decides the order in which the candidates appear on the party lists. This factor can of course be crucial.

Voting rights in Bavaria – the only federal state to do this – stipulate the use of so-called loose or open party lists[13]. It is for this reason, namely because the voters are offered greater scope in influencing the choice of the

---

[13] "Das bayerische Wahlrecht kennt, als einziges in der Bundesrepublik, lose gebundene Listen." Emil Hübner. Das bayerische Landtagswahlrecht *in* das Regierungssystem des Freistaates Bayern, Band 11. Vögel. p.281.

actual deputies who represent them, that the Bavarian electoral system has been described as being extremely democratic.

Whereas in a German federal election the second votes decide, by means of proportional representation, the strengths of the parties in parliament – and therefore the second vote is actually more important than the first – in a Bavarian state election both votes can be equally important in deciding which actual candidates will enter the *Landtag* in Munich, because the total number of votes a candidate receives from first votes (if he or she is a constituency candidate) and second votes is decisive. Not all candidates, of course, contest a constituency as well as appearing on a party list.

However, this system clearly gives an advantage to candidates who are both constituency and list candidates, and especially to candidates in more densely populated constituencies, for example in Munich or other urban areas, where a larger number of votes can be recorded than in some of the rural constituencies.

Supporters of the Bavarian electoral system argue that this emphasis on a closer relationship between the electorate and its candidates is a positive feature, since research in the early years of the Federal Republic revealed that most German voters cast their vote for a party rather than a particular candidate. Studies of how little voters knew about constituency candidates in federal elections confirm this, and there is little evidence to show that things have changed significantly[14].

Additional seats

It is, therefore, the total of the first and second votes that determines the order in which the mandates are allocated. Constituency candidates normally enter parliament of course, except in the rare case of an additional seat (*Überhangmandat*).

At federal elections in Germany only a small number of such additional seats occurred in most cases. At some federal elections no such extra mandates were created. However, there were six extra seats at the first all-German federal election in 1990, because the CDU won every constituency seat in Saxony and Thuringia, and all but one in Mecklenburg-

---

[14] Hans Apel. Der deutsche Parlamentarismuus. Rowohlt. Hamburg. 1968.

Vorpommern. At the second federal elections in the new Germany in 1994 a record number of no less than sixteen additional seats occurred (twelve for the CDU and four for the SPD) – somewhat controversially in fact, in a situation where the government coalition had a majority of only ten mandates. In such cases, where a party gains more constituency seats (from the first votes) than it is strictly speaking entitled to on the basis of proportional representation (from the second votes), the number of seats in parliament in Bonn is increased.

The occurrence of the six additional seats meant that from 1990-1994 there were 662 German members of parliament in Bonn instead of the envisaged 656, the new figure after German Unity. After the October 1994 election the Bundestag contained 672 members, making it the largest parliament of any country with free elections. Initially Bavarian electoral law also recognised additional seats, but not an increase in the size of parliament. The present position is that *Überhangmandate*, should they occur, are permitted and the size of parliament would be increased accordingly. However, with the exception of 1950 and 1954, no additional seats have occurred in practice in Bavarian state elections.

Under Bavarian electoral law of 29.3.1949, obtaining for the 1950 state election, at which two additional seats were gained by the CSU, the seats were retained, but at the expense of the other parties. What actually happened in 1950 was that in Swabia 29 seats had to be allocated on the basis of applying the d'Hondt system of PR to the votes cast: the CSU gained ten seats, the SPD eight, the BP five, the BHE four and the FDP two. However, of the fifteen Swabian constituencies, decided on first votes, the CSU won twelve and the SPD three.

The two additional seats gained by the CSU were retained, not by increasing the number of seats for Swabia (as would happen nowadays in Bavaria or in a federal election), but at the expense of the other parties. This was done by allocating only seven, instead of eight seats to the SPD and only four, instead of five to the BP. In this way the total number of parliamentary seats in Swabia (and Bavaria) was not increased.[15]

---

[15] Figures taken from Die Landtagswahl von A bis Z 1986, published by das Bayerische Landesamt für Statistik und Datenverarbeitung, p.52.

In fact the two additional seats for the CSU in 1950 were the explanation of the fact that it had one more parliamentary seat than the SPD, even though the SPD polled exactly 28 per cent of the vote and the CSU polled 27.4 per cent in Bavaria as a whole. The law was changed on 22.12.1952, consequently valid for the 1954 Bavarian elections, at which there were again two *Überhangmandate*. On that occasion the CSU in Lower Bavaria gained two additional seats. On the basis of PR the 25 mandates in *Niederbayern* (Lower Bavaria) were allocated as follows: CSU – ten, BP – seven, SPD – five, BHE – three. The CSU, however, won all twelve of the constituencies, producing two additional seats.

Since the 1952 change in the electoral law disallowed additional seats, the two constituency candidates with the lowest number of votes – Herr Heigl and Dr Schmid – lost their constituency mandates. Ever since 1954 no additional seats have occurred in practice, although the electoral law of 27.7.1973 amended the position on additional seats, so that, had any occurred, the number of parliamentary seats in the Maximilianeum would have been increased.

With the sole exception of the very first election to the Munich parliament in 1946, when there 180 seats, there have been 204 parliamentary seats in Munich ever since. At the 1994 Bavarian election 104 seats were constituency mandates and 100 came from the party lists. After the 1994 federal election 89 Bavarian deputies were returned to the Bundestag in Bonn – 45 via constituencies and 44 via party lists.

Parties which in federal elections cannot gain five per cent may still be represented in the *Bundestag* if they win three constituency seats. There is no equivalent regulation in Bavarian electoral law.

Changes for 1994

In the spring of 1992 a number of articles in the Bavarian press referred to the fact that, strictly speaking, the Bavarian parliament, elected on 14 October 1990 was unconstitutional. Seventy-seven members of the Bavarian parliament from the FDP, SPD and the Greens complained to the Bavarian Constitutional Court, and in judgments on 24 April and 19 May 1992 their complaint was upheld.

Leo Parsch, the president of Bavaria's highest court was quoted as saying that, under the system of counting the votes separately in each of the seven Bavarian districts and not carrying over remaining votes but simply discarding them, the small parties were disadvantaged too much.16

Max Stadler, the chairman of the Bavarian FDP, calculated that his party needed approximately 82,000 votes on average to win a mandate in 1990, whilst the CSU needed only 48,000 per mandate. The Bavarian liberal party successfully appealed to the Bavarian Constitutional Court in 1992. Although the decision of the court meant that between 1990 and 1994 one SPD and six CSU politicians had entered parliament unconstitutionally, the Court decided that no action would be taken. Nevertheless the court decided that the seat allocation after the Bavarian state elections on 25 September 1994 would take place acording to the Hare-Niemeyer method (named after two mathematicians, one British and one German). This proportional method of seat distribution replaced the traditional d'Hondt method (named after a Belgian mathematician and lawyer, Victor d'Hondt) for federal elections from 1987 onwards. The Hare-Niemeyer system favours smaller parties slightly more than the d'Hondt system. The Bavarian FDP first protested vigorously over the discrepancy between the percentage of votes and seats gained – the system always benefitted the large parties and disadvantaged the smaller ones – in 1978.

It was estimated that the state elections in Bavaria 1978, whilst the CSU and the SPD required around 50,000 to 52,000 votes in order to gain a mandate, the FDP required some 70,000[17]. The FDP had demanded a revision of the system of seat allocation and disregarded votes even then.

In 1990 the CSU gained 54.9 per cent of the votes but, with 127 of the 204 mandates, 62.3 per cent of the seats. The SPD gained 26 per cent of the votes but 28.4 per cent of the seats. The Greens, with 6.4 per cent gained only 5.9 per cent of the seats (12 mandates) and the FDP with 5.2 per cent were allocated 3.4 per cent of the seats in parliament (7 mandates). If the Hare-Niemeyer system of distribution had been used the following number of seats would have emerged: CSU – 121, SPD – 57, Greens – 15, FDP – 11.

It should of course also be remembered that the electoral picture is also distorted slightly by the vote received by parties which do not clear the five

[16] See the Munich AZ 25.4.1992, p.5.

[17] FDP. Die Liberalen. Arbeitsprogramm 1979-82. p.11.

per cent hurdle. This accounted in 1990, for example, for 7.5 per cent of the vote, including the Republicans (4.9 per cent) and the ÖDP ecology party with 1.7 per cent. The seats in parliament are of course allocated as if those parties which do win seats had gained one hundred per cent of the vote between them. This is never the case.

There has however, as yet, been no call to abolish the five per cent clause, since that would mark a return to the purely proportional days of the Weimar Republic, which proved unmanageable. Although the Bavarian electoral system is basically one of proportional representation – it therefore produces far less variance between the proportion of electoral support a party receives and the proportion of parliamentary seats it is awarded than is the case in Britain or the USA – political parties in Bavaria, especially the smaller ones, would like, and indeed have demanded, as "fair" a system as it is possible to get.

For this reason the author was told in autumn 1993, by the Head of the Bavarian FDP association in Munich[18], that the Bavarian liberal party was proud to have been the driving force behind a small but significant victory in the area of electoral reform.

Local elections in Bavaria

Electoral law for local elections (*Kommunalwahlen*), which since 1960 have been held every six years in Bavaria, is even more "democratic" and complicated. It gives the voter the opportunity of "accumulating" or casting up to three votes for one candidate, if he or she wishes (*kumulieren*). No five per cent clause operates in local elections.

The number of votes which each voter has varies and depends on the number of candidates standing. At the local elections on 18.3.1990 in Creußen for example, there were sixteen candidates for each of six parties[19]. The voter therefore had sixteen votes. Each voter was allowed to give either one, two or three (but no more than three) votes to any one candidate and no more than sixteen votes in total. In Munich there were eighty seats up for election on the Stadtrat (town/city council), so each voter had eighty

[18] Interview with Herr Hans-Helmu Rösler in the headquarters of the FDP *Landesverband* in Munich on 20 October 1993.

[19] See op. cit. Roth. Freistaat Bayern. Politische Landeskunde. p.67

votes. In Memmingen there were forty.

As the voter may also distribute the votes at random across party lists (known as *panaschieren*), he or she is not restricted to giving votes only to candidates from the same party. Investigations show that Bavarian voters do in fact make considerable use of both these possibilities: *kumulieren* and *panaschieren*.

This does, however, mean that the ballot papers are often the size of a broadsheet newspaper and that the electoral system applied to local elections in Bavaria is complicated. Some researchers have claimed that the system is not understood properly by all the voters. The final details of all the results of local polls are not usually available until about two months after the election. The complete final results of the March 1990 Bavarian local elections were not available until the end of May of that year.

The Bavarian constitution underlines, as perhaps is only to be expected, the special stance which Bavaria so often adopts. It displays the hallmarks of a constitution for a Land which in 1946 was not really a new creation, given Bavaria's long and enduring traditions. It was more a case of designing a constitution for a Land which was being re-constituted, rather than being freshly created. It must also be viewed in the context of a new constitution which was approved nearly two-and-a-half years before the Federal Republic and its provisional constitution, the Basic Law existed. It remained to be seen which political parties would be established in postwar Bavaria, and how they could adapt to the new constitution.

5

Parties and Elections in Postwar Bavaria

Parties and elections in post-war Bavarian politics have been dominated by the CSU, the Christian Social Union party (*die Christlich-Soziale Union*). Although it was a new party, founded in 1945 and licensed by the US occupation forces as a *Land* party early in 1946, the CSU skilfully built upon the strong historical and political traditions of Bavaria's long and eventful past. Since the hegemony and remarkable achievements of the CSU require lengthy and detailed treatment, let us first look at the overall picture and the other parties involved in postwar Bavarian politics.

The genesis of a new Bavarian party system

The American occupation forces exercised great caution in scrutinising the re-establishment of political life, showing great suspicion of any organisation which might have sympathies with the NSDAP. This sometimes led to misunderstandings by American commanders, whose limited knowledge of German made them suspicious of any mention of the German word "rechts" (right). To the amusement of some Germans, several of the American commanders made the false assumption that a *Völkerrechtspartei* (legally constituted party) was a right-wing Nazi party.

Political parties in Bavaria were first permitted at local level on

27 August 1945, at Land level, i.e. for the whole of Bavaria, on 23 November 1945 and at zone level, i.e. for the US, British or French Occupation Zone in western Germany, from February 1946.

Initially things began hesitantly. Many people were suffering from shock, confusion and disorientation after German capitulation and collapse. By around the end of November 1945 sixty-three local political organisations had been constituted in Bavaria. Of these 23 belonged to the SPD, 21 to the KPD, and the remaining 19 represented nine different political directions.[1]

The first political parties to be granted a licence in the re-constituted *Land* of Bavaria were the SPD, the CSU, the Communist KPD, the Economic Reconstruction Party WAV (*die Wirtschaftliche Aufbauvereinigung*) and the liberal FDP. As these five were the only parties initially, the Bavarian party system began to develop along the lines of a four-strand model: Christian/conservative – Liberal – Socialist – Communist. These four broad strands had also been clearly visible in earlier Bavarian party systems.

New parties for a new political system

WAV

The WAV, or Economic Reconstruction Party, existed only in Bavaria. The party was founded by Alfred Loritz, a Munich lawyer who gained some support from business circles and also amongst refugees. After 1945 approximately two million refugees, mainly from Czechoslovakia and Silesia, settled in Bavaria. The Germans from the Sudetenland in Czechoslovakia, the *Sudetendeutsche*, were the largest single group of refugees (around one million) in Bavaria, and there were about half a million Silesians.

At a meeting on 26 January 1956 Dr Wilhelm Hoegner (SPD), the Bavarian minister president at the time, greeted the Sudetendeutsche as Bavaria's fourth people or "tribe" (der vierte Stamm), after the Old Bavarians, the Franconians and the Swabians. Even today regular (biennial) meetings are still held of the Sudetenland Association in Munich[2].

---

[1] Nürnberger Nachrichten 27.11.1945.p.2.

[2] This meeting in 1956 is referred to in an article by Heinrich Kuhn in a special supplement on Bavaria, das Parlament no. 11, 19. 3. 1983, pp.6,7.

At the constitutional assembly elections on 30 June 1946 the right-of-centre WAV polled 5.1 per cent and at the first Bavarian state elections on 1 December 1946 7.4 per cent (225,000 votes). It took part in the first Bavarian coalition government with the CSU and the SPD under Hans Ehard, but the parliamentary party collapsed on 17 March 1949.

At the next state elections in 1950 the WAV gained only 2.8 per cent of the Bavarian vote. It combined its efforts with another party, the "association of new citizens" (*Neubürgerbund*) for the 1949 federal elections and managed to record 14.4 per cent of the vote in Bavaria, gaining twelve mandates. By 1951, however, the WAV disappeared completely, after having briefly provided the Bavarian party system with an extra dimension.

KPD

The Communist Party, the KPD (*Kommunistische Partei Deutschlands*) was the first party allowed to re-form in postwar Bavaria. The American occupation forces actually insisted that the very first Bavarian cabinet (28.9.1941 – 21.12.1946) led by the minister president they appointed, Wilhelm Hoegner, included Communists. Heinrich Schmitt from the KPD became minister with special responsibility for de-nazification and there were two Communist secretaries of state, one at the Bavarian Ministry of the Interior and one in the Economics Ministry.

Initially, of course, having been a Communist during the Third Reich was at least proof for the US occupation forces that you were not a Nazi. In spite of this "positive discrimination" at the beginning, Bavaria's past political traditions did not favour the firm establishment of a Communist Party.

Despite the short-lived Socialist Republic established in Munich based on the Soviet model in 1918 (see chapter two) – or perhaps precisely because of it – the political Left had little appeal in Bavaria, and even less as relations between the US and the USSR worsened in the early postwar years. The forced marriage of the SPD and the KPD in April 1946 to form the East German SED (*Sozialistische Einheitspartei Deutschlands*), plus the later establishment of the German Democratic Republic, in October 1949, had the effect of decreasing even further the limited appeal the KPD had amongst the West German electorate in general and the Bavarian electorate in particular.

In Bavaria in 1946 the KPD gained 5.3 per cent in the elections to the Constitutional Assembly (nine seats), but never gained any representation in the Bavarian parliament, since they never succeeded in clearing the ten per cent hurdle. The author was told in several interviews[3] in Munich that one of the original reasons for the Bavarian ten per cent hurdle was the exclusion of the Communist Party.

Its vote of 6.1 per cent at the first Landtag elections at the end of 1946 dropped to around two per cent before the Federal Constitutional Court in Karlsruhe banned the KPD in West Germany on 17. August 1956. This ban, which lasted until 1968, was controversial at the time because many people felt it was unnecessary, since the Communists were a rapidly declining political force anyway. Even when the party just cleared the Federal Republic's five per cent clause in 1949 (5.7 per cent), its result in Bavaria was below average (4.1 per cent).

SPD

The Bavarian SPD has always been very different from the federal Social Democratic Party and its associations in other regions of Germany. Soon after German unification in 1871 the SPD, the oldest of Germany's major parties, was referred to in Bavaria as the "monarchist" or "royal" branch of the party: *die königlich bayerische Sozialdemokratie*. This term is still heard occasionally nowadays, as a way of emphasising that the SPD in Bavaria is in a category of its own. A number of factors have produced a situation whereby the SPD has performed worse in Bavaria since the war than in any other *Land*. In fact the SPD polls around ten per cent less in Bavaria in both federal and state elections.

This state of affairs is partly the result of CSU hegemony (see below), but partly also due to the fact that the circumstances which obtained in Bavaria in terms of denominational, socio-economic and cultural factors

---

[3] This point was made in an interview with Theo Barfßfl, the former Reichsvorsitzender der Jungdemokraten, in Munich on 23.6.1982. It was explained to the author that in the discussions on the new constitution, political and electoral systems for Bavaria in 1945/46, there was a strong desire to exclude the Communists as far as possible.

were unfavourable for the social democrats.

The Bavarian SPD therefore has little in common with its more successful counterparts in other *Länder* (e.g. the SPD in Bremen or Hamburg). The Bavarian SPD was certainly seen as more of a people's party (*Volkspartei*) than a class party in the period 1945-59[4], even before the fundamental change brought by the Godesberg Programme in 1959 for the federal SPD. As Professor Paterson points out, the social democrats in Bavaria have suffered from the traditional handicaps of a secular and un-Bavarian identity, as well as factional strife when the SPD in Munich, which has one of the largest concentrations of students in the FRG, was taken over by the far left of the party.[5]

FDP

The liberal Free Democratic Party (*Freie Demokratische Partei*) was the last of the initial group of five parties (KPD, SPD, WAV, CSU, FDP) to be founded in post-war Bavaria. In that first phase, the Bavarian liberal party suffered from the fact that it was not founded as a *Landespartei* until 15 May 1946, only two weeks before the first important *Land* elections in Bavaria for the Constitutional Assembly on 30 June 1946. This was certainly one reason why the Free Democrats did so badly, polling only 2.5 per cent and gaining only three of the 180 seats in the assembly.

The FDP in Bavaria has never been able to adopt its federal role of majority maker or ideological balancer in the Bavarian party system, since its room for manoeuvre has always been much more restricted. At first the Bavarian Liberals benefitted from having Thomas Dehler as the chairman of both the Bavarian and the federal FDP parties. Dehler was considered one of the "fathers" of the Basic Law, and as, Federal Justice Minister, he was a member of Adenauer's first cabinet in 1949.

The particular traditions and political culture of Bavaria were however not the ideal breeding ground for the promotion of political liberalism.

---

[4] . . . kann man sagen, daß in Bayern die SPD mehr eine Volkspartei als eine Klassenpartei ist, was wiederum aus der Struktur des Landes erklärlich ist." Ilse Unger. Die Bayernpartei. Geschichte und Struktur 1945-1957. dva. Stuttgart. 1979 p.59.

[5] William E. Paterson and David Southern. Governing Germany. Blackwell. Oxford. 1991. p.196.

"Liberalitas Bavarica," the inscription over the church door in the monastery village of Polling, near Weilheim in Upper Bavaria, has a deeply traditional, ultra-conservative sense of Bavarian freedom and tolerance of religion, which according to Wilhelm Hausenstein meant that an open religious approach must be capable of accepting new influences without losing or renouncing its own. A "liberal" view in that particular Bavarian context was concerned more with preserving the status quo than any sort of modern definitions of progressive and dynamic political liberalism.

Things are different in Bavaria, and the pivotal role played by the liberal party at federal level in Bonn via the device of coalition politics has simply not been on offer in Munich. Indeed, over the last fifteen years, it has been even more of a struggle for the FDP to clear the amended five per cent electoral hurdle in Munich than in Bonn.

Consequently the FDP has consistently recorded its worst election results in Bavaria, averaging only around six per cent in state elections there. On three occasions since 1946 the FDP has failed to enter parliament in Bavaria: once it failed to clear the Bavarian ten per cent hurdle (in 1966), and on three more recent occasions the Bavarian Liberals fell foul of the five per cent clause, including the 1994 Bavarian state election when the FDP polled its worst-ever result of 2.8 per cent. In German federal elections the party has averaged just over nine per cent overall, but only 6.6 per cent (lower than any other federal state) in Bavaria.

In his discussion of the development of political liberalism in Bavaria, Friedrich Henning, the former Head of the FDP archives in the Friedrich Neumann Foundation of the Thomas Dehler House in Bonn, emphasises three main points. Firstly, he speaks of the very small recruitment potential for the FDP in Bavaria, owing to the special structure of the region in terms of socio-economic and religious/denominational factors.

Secondly, he stresses that the power distribution in favour of the liberals' opponents was such that liberal groups were forced to join forces in nineteenth century Bavaria more than in other areas. This cooperation was not always successful and left the post-war FDP in a weak position. Thirdly, Henning draws attention to the static and conservative thinking which often influenced Bavarian governments and the constant uphill battle fought by the FDP and its predecessors against illiberal and at times immobile policies[6].

The CSU – Bavaria's most successful party

No party in the Federal Republic of Germany has exerted as much political power and influence and regularly recorded such spectacular election results as has been the case with the CSU in Bavaria. It has consistently out-performed both the CDU in any of the German Länder and the SPD in its strongholds. The CSU averaged 53 per cent at the federal elections between 1949 and 1990, and its average result in Bavarian state elections during the same period is just over 50 per cent.

Although the Christian Social Union was a new party, founded in 1945/46, it did re-establish a Christian/conservative, stronly Roman Catholic tradition in Bavaria. The predecessors of the CSU can be traced back to the Bavarian Patriot Party (*Bayerische Patriotische Partei*), which entered parliament in Bavaria in 1869 with over forty per cent of the vote. After a much smaller representation for the Bavarian Patriots in the parliament of 1881, Catholic interests were represented politically by the Centre Party (*Zentrum*) from 1887 onwards.

The Catholic Centre Party was influential in Germany in the late nineteenth and early twentieth century; it played an important part in the Bavarian parliament, where it was the largest party between 1887 and 1912, with over fifty per cent of the seats in Bavaria for most of that period. This established a clear trend in Bavarian voting patterns which continued after the 1914-18 war.

During the Weimar Republic political Catholicism in Bavaria made its presence felt in the shape of the Bavarian People's Party, the BVP (*Bayerische Volkspartei*). The BVP was formed from the Bavarian wing of the Catholic Centre Party and became the immediate predecessor of the CSU.

In his book on Bavarian and German federalism, Dorondo reaches the conclusion that although the BVP struggled hard to further Bavarian inter-

---

[6] Friedrich Henning. Entwicklungen des politischen Liberalismus in Bayern im 19. und 20. Jahrhundert. Nachwort zur Dissertation von Berthold Mauch: Die bayerische FDP. Portait einer Landesparatei 1945-49. Olzog. München. 1981, pp.97-109.

ests within the German framework of federalism, the party never succeeded in breaking out of the purely Catholic, middle class, (Old) Bavarian mould[7]. It also lost some votes to its competitor, the Bavarian Farmers' Association (*Bayerischer Bauernbund*), which represented Old Bavarian Catholic farming and agricultural interests. This split the Catholic vote and meant that, as it was incapable of appealing to a wider Bavarian constituency, the BVP was unable to break out of the "thirty per cent tower."

A new party is founded

The Christian Social Union was originally founded as a political party on the basis of local initiatives throughout Bavaria. For example, a local party was founded in Bamberg, and then for the whole of Upper Franconia, on 16 September 1945 by a group led by Dr Gerhard Kroll. In Nuremberg a CSU party was founded on 13 October on the initiative of a psychologist, Dr Wilhelm Arnold, on the Catholic side, and a lawyer, Dr Paul Nerreter, on the Protestant side.

Adam Stegerwald, a former Prussian minister president and *Reichsminister*, founded a branch of the CSU in his home town of Würzburg on 10 October 1945. This date is usually taken as the official founding of the CSU. Stegerwald, who died in December 1945, is considered a co-founder of the party.

On 25 November 1945 the influential Munich committee of the CSU met, calling for the formation of an inter-denominational party based on Christian principles and morality. This committee agreed on 17 December 1945 that a party should be established for the whole of Bavaria and that the Munich lawyer Josef Müller, another important co-founder of the party, should be the future chairman.

Nevertheless political parties had still not been licensed by the American military government on a *Land* basis. The question of names of parties was also a sensitive subject. At the inaugural meeting on 8 January 1946 in Munich's town hall several names were suggested for the party which was to be founded for the whole of Bavaria.

One popular proposed name, the Bavarian Christian Social Union, was

---

[7] D. R. Dorondo. Bavaria and German Federalism. Reich to Republic, 1918-33, 1945-49. St. Martin's Press. New York. 1992.

rejected by the Americans Occupation Forces, who did not wish to prejudice the chances of a future Germany, not yet re-established, so the first word was omitted. The word Christian was nearly omitted by the party's more progessive wing, but the name CSU was finally agreed and the US military government licensed the CSU on a *Land* basis on the same day: 8 January 1946.

From the very beginning there was a strong feeling amongst many of the founding members that the new party should be a Christian, collective movement, a people's party (*Volkspartei*) of mass appeal, which should aim at bringing together the Protestant and Catholic denominations.

Two wings of the party develop

However there were two distinct factions. One centred around the genial Josef Müller, known affectionately as der Ochsensepp (Sepp is the Bavarian abbreviation for Joe). He was a Munich lawyer with an impish sense of humour who came from Upper Franconia, and was, with Stegerwald, a co-founder of the party. The other, more conservative, traditionalist faction was basically the Catholic-Bavarian wing centred around Fritz Schäffer. These two wings have been referred to as new direction (*Neuorientierung*) versus tradition (*Tradition*).

The Müller wing of the party was the more progressive, liberal-conservative, inter-denominational group, which included people like Michael Horlacher, Wilhelm Eichhorn, Heinrich Krehle and Franz Steber. The more traditionalist, fiercely moral, Roman Catholic, ultra-conservative wing included Fritz Schäffer, Anton Pfeiffer, Walter von Miller, Josef Baumgartner and Carl Lacherbauer. Baumgartner was the CSU minister of agriculture in Hans Ehard's first and second Bavarian cabinets. He resigned his post in January 1948 and became Chairman of the BP. Lacherbauer also joined the BP.

Another key personality, Alois Hundhammer, belonged to the same wing of the CSU. Known at the time amongst his colleagues as "the Christian tank", Dr Dr[8] Hundhammer was described to the author in an interview in Bavaria as the epitomy of the ultra-conservative, clerical politi-

---

[8] In Germany it is possible to have the title Dr Dr, if you have written two PhD theses.

cian with a tendency to behave as if he were still living in the Middle Ages[9].

Hundhammer, who became Bavarian education minister in 1946/47, was fanatical in his defence of his moral and religious beliefs. In the Weimar Republic he had been the deputy General Secretary of the Bavarian Christian Farmers' Association and a member of parliament for the BVP. His nickname of "the Christian tank" was acquired because of his powerful crusades for the Christian cause, which were sometimes criticised for making the word Christian synonymous with the word Catholic. As education minister, he appeared obsessed with the idea that there was a movement trying to ban the cross from Bavarian classrooms.

If the liberal/progressive Müller wing was going to succeed in its ambitions of establishing a more open, inter-denominational party representing the interests of the whole of Bavaria, then it had to integrate the Franconian, Protestant, urban/industrial areas, whilst retaining the support of the farmers and Catholics in south and south-eastern Bavaria. Bavaria's north west corner, the district of Lower Franconia, with its capital Würzburg, is also mainly Catholic. Of the seven Bavarian districts, only two are predominantly Protestant: Upper and Central Franconia.

Table 4.1

Denominational breakdown within Bavaria

		% Catholics	% Protestants
Old	Oberbayern	75.1	17.6
Bavaria	Neiderbayern	90.1	7.1
	Oberpfalz	85.2	12.0
Franconia	Oberfranken	44.4	52.5
	Mittelfranken	35.9	58.5
	Unterfranken	75.0	20.5
Swabia	Schwaben	77.7	15.6

[9] "Dr Hundhammer war ein Prototyp des klerikalen-konservativen ins Mittelalter tendierten restaurativen Politikers." Interview with Dr Klaus Dehler, the nephew of the FDP Federal Justice minister in Adenauer's first cabinet, Thomas Dehler, conducted by the author in Nuremberg on 9.3.1983.

Fig 5.1 The Franconian Protestant "corridor" in Catholic Bavaria

Taken from:
A Mintzel, Die CSU 1945-72.
Westdeutscher Verlag.
Oplanden. 1975.

As can be seen from the table, even in the two districts where there are more Protestants than Catholics, the percentage of Protestants is nowhere near the very high figures for the number of Catholics in the other five districts.

The fraternal feud

Given the special set of circumstances which obtained in Bavaria, the CSU, only two years after being founded, had established itself as a party with great appeal to the native population (90.2 per cent Bavarian, 9.8 per cent non-Bavarian), and very much a Catholic party (91.3 Catholic, 8.4 per cent Protestant)[10], despite the aim of Müller and others to appeal to both denominations.

In the late Forties around two million (mainly Protestant) refugees and expellees settled in Bavaria; they were under-represented in the CSU, as were Protestants among the local population. As far as membership was concerned, the party's strongholds lay in the territory of Old Bavaria – the districts of the Upper Palatinate, Upper and Lower Bavaria – and Swabia. The four regional associations of the CSU in these districts had between them almost 69 per cent of the total membership, excluding the city associations of Munich and Augsburg. The three Franconian districts accounted for only a quarter of the CSU's 82,200 members at that time.

The Bavarian Party

The Bavarian Party (*Bayernpartei* – BP) was founded in Munich on 28 October 1946, but was allowed to operate only at local level, since the Occupation Forces in Western Germany were cautious of regional parties until a new German state was established. On 29 March 1948 the US military government in Bavaria permitted the Bavarian Party to be licensed at *Land* level. Until then only the CSU, SPD, KPD, FDP and WAV had been licensed. The BP was a radically federalist, strongly Catholic, ultra conservative party which took great pride in all Bavaria's past achievements. It was a loyal defender of rural Bavarian farming and agricultural interests.

The BP rapidly developed into a dangerous competitor for the CSU, as

---

[10] Alf Mintzel. Geschichte der CSU. Ein Überblick. Westdeutscher Verlag. Opladen. 1977, p.63.

substantial numbers of people transferred their membership from the CSU to the BP. The CSU, of course, wanted to establish itself as *the* Bavarian party, but it had not been allowed to include the word Bavarian in its party name in 1946. Many insults were hurled between the two parties, some of which did nothing to suggest that there was much love lost between the two Bavarian "brothers." On one occasion the CSU allegedly called the BP "Communists", not the act of a Christian party, according to the BP.

The currency reform on 20 June 1948 accelerated the loss of members in the CSU. As the former BVP traditionalist wing of the party, represented by Hundhammer and Schäffer, gained the upper hand, Josef Müller's vision of an inter-denominational Christian mass movement party with a broad organisational base, acting as a realistic political alternative to the SPD, began to fade. CSU politicians, arguing amongst themselves, seemed at that point in time to be incapable of getting to grips with the strong socio-political barriers and established cultural patterns in Bavaria.

The BP was founded by Ludwig Lallinger, a Munich policeman. It claimed to represent the political aims of all those who wished to promote the special rights, status and identity of Bavaria. Its slogan was Bavaria for the Bavarians (*Bayern den Bayern*). It first contested the local elections in Bavaria in 1948, where it achieved some notable successes in various town councils and rural boroughs.

Several dissatisfied former CSU members joined the Bavarian Party, as well as supporters of the Bavarian Democratic Union, which demanded an autonomous Bavaria. Another party, the *Bayerische Heimat- und Königspartei*, a sort of Bavarian monarchist group, was founded in 1947, but banned by the US Occupation Forces. The majority of its members, many of whom came from the Chiemgau area, transferred their membership to the BP.

The Bavarian Party gained most of its support from the predominantly Catholic districts of Lower and Upper Bavaria; it managed to make a big dent in the CSU vote of 52.3 per cent in 1946 (by 1950 the CSU vote dropped to 27.4 per cent) by polling 17.9 per cent of the vote – and gaining 39 seats – in 1950. The BP attracted 13.2 per cent (28 seats) at the 1954 Bavarian state elections. That level of support dropped at the next two state elections (*Landtagswahlen*) in 1958 and 1962 to 8.1 (14 seats) and 4.8 per cent respectively. Even this last figure brought the party eight deputies in the

Bavarian parliament, since the BP cleared the Bavarian ten per cent hurdle, as it was then, in its stronghold of Lower Bavaria with 10.3 per cent.

After one of the BP's main politicians in the CSU-led Bavarian cabinet, Robert Wehgartner, transferred his allegiances to the CSU in 1966, the BP disappeared from an already weak position into virtual oblivion.

The BP, however, did cause a stir at the Federal Republic's first federal elections when it polled over a fifth of the Bavarian vote (20.9 per cent – this represented 4.2 per cent of the West German vote), securing 17 seats in the Bundestag. In 1949, remember, the West German voter had only one vote in national elections and the electoral system required parties to gain at least five per cent of the valid second votes in any *Land*.

In 1953, when the federal electoral hurdle was tightened to five per cent in the FRG, the BP, with 9.2 per cent in Bavaria which averaged out to only 1.7 per cent nationally failed to enter parliament in Bonn. The BP of course did not campaign outside Bavaria. In 1957 the BP joined with the Zentrum to form the FU (*Föderalistische Union*) for the federal elections. It managed only 3.2 per cent of the vote. Professor Gordon Smith points out that it is difficult to avoid the impression that the change in the electoral law implemented for the 1953 election[11] was deliberately aimed at excluding the Bavarian Party.

The Party of Refugees and Expellees (BHE)

The party with one of the longest and most awkward names in post-war Germany was the BHE, *Block der Heimatvertriebenen und Entrechteten* – the league of those expelled from their homeland and those deprived of their rights. Initially the party had some electoral success in Schleswig-Holstein and Lower Saxony; it campaigned in several areas of the newly founded Federal Republic, which contained around twelve million refugees of one sort or another. Almost two million of these were to be found in Bavaria. The BHE added and subtracted various initials to its name from time to time.

The BHE took part in the 1948 local elections in Bavaria and in July 1950, when the US authorities granted them a licence, a Bavarian *Land*

---

[11] Gordon Smith. Democracy in Western Germany. Heinemann. London. 1979, p.110.

party was formed. In 1952 the Bavarian branch of the BHE, like those in other regions, added the name *Gesamt-deutscher Block* to its already clumsy title, in an attempt to attract more votes. The GB/BHE did succeed in absorbing most of the WAV's former supporters, since the WAV was not represented in the second Munich parliament.

On 26 November 1950 the BHE received 12.3 per cent of the vote in Bavaria and was allocated 26 seats. At that time the party was in alliance with another small party known as the *Deutsche Gemeinschaft* (DG). Four years later, at the next Bavarian elections on 28 November 1954, the GB/BHE gained 10.2 per cent (19 seats) and was one of the government parties in the Bavarian Coalition of Four – *die Viererkoalition* – along with the SPD, FDP and the BP. That was the only Bavarian government since 1945, right up to the present day, which has not included the CSU. It lasted only three years, from 1954 until 1957.

As Germany's refugees became more integrated into Bavarian (and German) society, the appeal of the party gradually diminished. By 1958 its support in Bavaria had dropped to 8.6 per cent and only 17 seats. Although the BHE still managed 5.1 per cent at the 1962 state elections, it failed to clear the ten per cent hurdle in any one of Bavaria's government districts.

The only federal election at which the BHE gained any mandates was the second one in 1953, at which it secured 8 seats. One of the party's leading national politicians, Frank Seiboth, wanted to unite two other small parties, the FDP and the German Party (*Deutsche Partei* – DP) for the 1957 election to the Bundestag, but the attempt failed. Within Bavaria, the BHE consistently derived most of its support from Lower Bavaria and parts of Swabia and Upper Franconia.

The neo-Nazi NPD

The National Democratic Party of Germany (*die Nationaldemokratische Partei Deutschlands*), the NPD, founded in Hanover on 28 November 1964, played a brief but well publicised part in the Bavarian party system from 1966 until 1970, when it replaced the FDP as the third party in the *Maximilianeum*, the impressive building in Munich, which houses Bavaria's parliament or *Landtag*.

On 20 November 1966 the NPD gained 7.4 per cent overall in Bavaria, but managed to clear the ten per cent hurdle in Central Franconia with 12.2 per cent which was precisely the district where the Bavarian Liberals (FDP 1966: 9.0 per cent) usually did so. The NPD gained 15 deputies in the Bavarian parliament. The strongholds of the thinly disguised German neo-Nazi party at that time were Central Franconia, northern Hesse and Lower Saxony, which corresponded fairly closely to those of its predecessor, the NSDAP. Its main appeal was in predominantly Protestant, middle-class farming areas[12].

During the Federal Republic's first major economic recession since the German economic miracle (*Wirtschaftswunder*), the NPD entered no less than seven *Land* parliaments and many town councils between 1966 and 1968. However the NPD's political life was short. The 781,813 votes it received at the 1966 Bavarian election dwindled to 162,823 (2.9 per cent) by 1970. Even though the NPD just cleared the five-per-cent-hurdle in Bavaria (5.3) and also in several other German states at the 1969 federal election, it gained only 4.3 per cent overall and therefore did not enter the *Bundestag* in Bonn.

A new phenomenon: the Greens

The formation of the Green Party (*Grüne Partei*) in Bavaria sprung from the AUD (*Aktionsgemeinschaft Unabhängiger Deutscher*). This was an active group of independent Germans, first founded in 1965, which itself had grown out of several citizens' initiative groups (*Bürgerinitiativen*) and a collection of different movements. These even included a group which claimed to represent an association of farmers and middle class citizens (*Bauern- und Mittelstandsbund*). The AUD later became a model for the federal Green party which contested the 1979 European elections.

The Bavarian Greens, like other German Green parties, stressed four key principles: the ecology, social aspects, root-and-branch democracy and non-violence: *ökologisch, sozial, basisdemokratisch, gewaltfrei*. However, it would be wrong to identify the ecology movement in Bavaria entirely with the Greens. They should be seen more as a protest movement against what

---

[12] Reinhol Kühnl Die NPD. Struktur, Programm und Ideologie einer neofaschistischen Partei. Voltaire Verlag. Berlin. 1967, p.58.

Alf Mintzel calls the "unholy trinity" of State (Bavaria), State Party (CSU) and the (Catholic) Church – *Staat, Staatspartei und katholische Kirche*[13].

When the Greens entered the Bavarian parliament for the first time in 1986, as the third party – they had been the third political force since 1984 – they tried to create a youthful and viable counter-cuture to the political hegemony of the CSU. The Greens and their numerous related groups were particularly active in the urban conurbations of Munich, Nuremberg/ Fürth, Augsburg, Aschaffenburg and Neu Ulm – in other words centrally located areas with associated service and supply industries.

Parts of Bavaria where rapid demographic and industrial development had occurred often had special environmental concerns, e.g. waste disposal, atomic power production, motorway construction and the like. In such areas the Bavarian Greens received at least modest support, especially from younger voters.

It is in precisely such developments that it can be seen that nowadays even Bavaria, with its special and enduring traditions and unique political culture, has been affected to a certain extent by some of the cultural and socio-economic changes which have made their mark on the Federal Republic as a whole.

Even the Free State of Bavaria has not been entirely immune to the general secularisation process in Germany, which has manifested itself in decreasing regular church attendance, an increase in the number of divorces and abortions and fewer children per family, as so-called "industrial" and "economic" values have come to the fore. Such factors have affected Bavarian society too, although to a lesser extent than in some other regions of Germany, given that Bavaria, even today, is still influenced by certain idiosynchratic aspects of its make-up, for example a high number of practising Catholics, and a relatively lower number of trade union members.

The late Petra Kelly's presence in Bavaria gave the Greens a slight boost – although she was very much a national figure – in an area where ecological issues and nature preservation have always been recognised as a

---

[13] "Das dreieinige Machtkartell in Bayern auch nach Strauß von Staat, Staatspartei und katholischer Kirche." Mintzel refers to this in F.J. Strauß. Der Charakter und die Masken. Der Progressive und der Konservative. Der Weltmann und der Hinterwäldler by Hans-Jürgen Heinrichs (ed.). Frankfurt am Main. Athenäum. 1989, p.172/173.

priority anyway. On the other hand however, conservative-minded Bavaria, where the CSU already played a dominant role before the rise of the Greens, was certainly not the ideal breeding ground for an anti-system party. Consequently the Greens did not come to prominence in Bavaria as much, or as soon, as in some other areas of Germany, e.g. Bremen, Hamburg, Baden-Württemberg.

The special circumstances of the party system obtaining in Germany's largest federal state mean that the Greens are confronted with an integrated, well-established, conservative environment, which permits a very restricted field in which to operate. In addition the Social Democrats are a much weaker political force in Bavaria than any other *Land*. In Bavaria at present there is no evidence of any SPD/Green co-operation, as there has been in neighbouring Hesse.

In later years there were controversial issues, to which the Bavarian Greens – whose majority faction consists of "Realos" – and others, turned their attention. For example, between 1983 and 1987 great controversy arose over the site of the nuclear waste reprocessing plant in Wackersdorf, near Schwandorf, about twenty-five miles north of Regensburg, in the Upper Palatinate. The Chernobyl nuclear disaster also affected Bavaria more than any other federal state. Although this issue was obviously taken up by the Green Party nationally, it brought no great benefit to the Bavarian Greens.

The strength of the West German Greens was often considered to be the way they had developed from grass-roots level. They styled themselves as the anti-party party (*"Wir sind die anti-Parteien Partei"*), and gained considerable support by dealing with local, mundane issues which affected people's everyday lives.

There was less scope for this in Bavaria, where the ruling party had a tighter grip on the day-to-day running of political affairs than in some other regions. The CSU took some of the ground from under the Greens' feet when in 1970 it established the first ministry for matters of the environment. Even though it also dealt with the development of Bavaria's territory (*Landesentwicklung*), it contained the key word *Umwelt* (environment) in its title. The CSU claim that Max Streibl in Bavaria, not Joschka Fischer in Hesse, was Germany's first minister of the environment (*Umweltminister*).

Some Green parties were successful at *Land* level before the formation of a political party at federal level in the 1979/80 period, or before the Green Party first entered the *Bundestag* in 1983, since it was a grass-roots party. Although the Bavarian Greens contested the 1978 state elections (1.8 per cent) – just after the party had been founded – they could not quite clear five per cent (4.6) in 1982, and it was not until 1986 that they first entered parliament in Bavaria, with 7.5 per cent of the vote. At that election the Greens took some 70,000 votes from the Bavarian SPD.

When the Green Party entered the Maximilianeum as third party – the Bavarian FDP polled less than five per cent in 1982 and 1986 – it was part of a trend. In 1986 the Greens were represented in seven out of the eleven West German state parliaments. Unfortunately, from their stand-point, their room for manouevre within the Bavarian political system, as has been demonstrated, was extremely limited, given the hegemony of the CSU and the frailty of the SPD in Bavaria.

Eight of the Greens' fifteen Bavarian mandates in 1986 went to women, which increased the total number of women in the Bavarian parliament to 25 out of 204. At that time that was the largest proportion of female deputies in Munich since 1946. In 1990 they again cleared the five-per-cent-hurdle (6.4 per cent), gaining twelve seats. The Green Party has a low average age (under 40), compared to the other parties in Bavaria, and a higher proportion of women candidates.

At federal level, a major achievement was the founding in 1993 of a new joint party *Bündnis 90/die Grünen*, which united all the Greens in east and west. The lack of a proper alliance between Greens in east and west Germany for the 1990 federal election was undoubtedly a contributory factor in the failure of the West German Greens to enter parliament in Bonn. Opinion poll surveys in early 1994 indicated around eight or nine per cent support for Bündnis 90/Grüne[14], the same as th FDP.

As the parties entered the "super election year" 1994, the spokesman for the Bavarian Greens, Gerard Häfner, said at the traditional Bavarian Ash Wednesday meeting (*politischer Aschermittwoch*) in February that his

---

[14] As reported on the Politbarometer on 25 February 1994 (a monthly survey undertaken by the Forschungsgruppe Wahlen in Mannheim for the German television channel ZDF).

party intended to fight hard in 1994 to break the CSU *Vollmacht* position; his party would not, he emphasised, consider a coalition with either the CSU or the Republicans.

Pressure from the right: the Republicans

The concomitant impotence of the opposition parties in Bavaria since 1970 – the CSU has had an absolute majority of both votes and seats at every Bavarian state election ever since that date – was threatened for the first time at the 1989 European elections by a new party, the Republicans (*die Republikaner*), who won 14.6 per cent of the vote in Bavaria (7.1 per cent in Germany), forcing *the* Bavarian party of state, the omnipotent CSU, down below the psychologically significant fifty per cent threshold. Bavaria was the only *Land* where the party's result entered double figures. At the 1989 elections to the European parliament 37 per cent of the Republicans' total vote in Germany came from Bavaria alone.

Even if some Bavarians would rather deny it, there can be no doubt that the Republicans (REPS) are essentially a Bavarian phenomenon, despite their successes in other regions too. In a 1993 publication on the extreme Right in Germany, Backes and Moreau describe the REPS as "a Bavarian product."[15] in its origins. The party was founded in Munich in November 1983 by three former CSU politicians: Voigt, Handlos and Schönhuber.

It is generally assumed that it was the arrangement of a massive loan (*Milliardenkredit*) for Honecker's GDR by Franz Josef Strauß which sparked off the bitter rows between FJS and his three former CSU colleagues and the subsequent founding of the Republican Party. Although that incident was certainly a factor, the author was told in an interview at the CSU headquarters in Munich in 1993[16] that in reality it was Schönhuber's 1981 book "*Ich war dabei*" (I was part of it) and his dismissal in April 1982 from his job as a presenter on Bavarian television which was the real catalyst in causing the rift.

---

[15] Uwe Backes and Patrick Moreau. Die extreme Rechte in Deutschland. Akademischer Verlag. München. 1993, p.51: Vergegenwärtigt man sich die Ursprünge der Partei "Die Republikaner" (REP), handelt es sich um ein bayerisches Entwicklungsprodukt.

[16] Interview with Dr Guber in Franz Josef Straß Haus, Nymphenburger Straße, Munich, on 15 October 1993.

It must be said, however, that the actions of Strauß in 1983 certainly met with widespread amazement in the CSU. In the light of subsequent events, i.e. reunification, it has been suggested by his most ardent supporters that maybe FJS understood the need to support the ailing GDR economy at the time, in order to support the East German regime until real change could be brought about in an organised manner.

In his controversial book Schönhuber, born in 1923, made it clear that he was proud of his membership of the Waffen SS and stressed the positive aspects of National Socialism, as he saw them. This tarnished his image as one of the most popular television personalities in Bavaria, having made a name for himself on the programme "Jetzt red' i" (now it's *my* turn to speak). On the programme, which is still running on Bavarian television, but with a different presenter, Schönhuber displayed great skill in helping ordinary Bavarian citizens, who had usually never appeared in front of a camera before, to relax and have their say on issues of the day. He derived great benefit from his populist appeal when the new party was founded; his confident media manner and public oratory skills were well received in the Bavarian beer cellars.

The first Republican party programme, passed by the "federal congress" in Munich on 26 November 1983, made a strong plea for the reunification of Germany. It bore the handwriting of Voigt and Handlos, who both left the party, after considerable internal friction, in 1986. Handlos joined the newly founded *Freiheitliche Volkspartei* (FVP), but with little success. Voigt turned briefly to the FDP but re-joined the REPS in 1989.

The Republicans' first electoral success was a vote of three per cent at the 1986 Bavarian state election, against all predictions. This brought the party DM 1.28 million in reimbursement of election expenses. The REPS sometimes close their election rallies with the German national anthem and some of their supporters cause friction by singing the now defunct first and second verses, still associated in many people's minds with the National Socialists.[17]

---

[17] "Deutschland, Deutschland über alles" was of course originally intended as a call to unite the many German states when it was written as part of a poem in 1841 by Hoffmann von Fallersleben. Because of the way it was widely misunderstood abroad, especially after its association with the Nazi party, the German national anthem now officially consists of the third verse only (Einigkeit und Recht und Freiheit).

The CSU "upstaged" the Republicans in 1986 by finishing their final election rally with the ubiquitous Bavarian brass band (*Blaskapelle*) playing the *Bayernhymne*, the Bavarian anthem. This song still has great emotional appeal for many of the inhabitants of Old Bavaria.[18]

A real breakthrough for the Republican Party came when it polled the unexpectedly high figure of 7.5 per cent in Janury 1989 in Berlin, where some parts of the city have very substantial Turkish populations. It was another controversial election campaign by the Republican Party. On that occasion great offence was taken by a television advertisement for the REPS, in which Turkish children were shown playing on the street in Berlin, with a song mentioning death in the background (*Spiel mir das Lied vom Tod*). This was of course assumed to be further evidence of the REPS's anti-foreigner policies (*Ausländerfeindlichkeit*) by their opponents. There was a certain irony in the fact that Schönhuber owned a holiday villa in Turkey at the time.

Although the Republican Party just failed to enter parliament in Bavaria in 1990 with 4.9 per cent, it gained a spectacular 10.9 per cent and 15 mandates in Baden-Württemberg in 1992, where, incidentally, both the major parties refused to form a coalition government with Schönhuber's party. By December 1992 membership had risen from the original 150, on the day when the party was founded, to 23,000.

Implications for Bavaria

Gordon Smith, commenting on the new German party system and the radical Right, makes the point that the CDU/CSU is particularly vulnerable to the appeal of parties such as the *Republikaner*. Professor Smith refers to the way such radical right-wing parties have frequently harnessed specific discontents, e.g. the size of the immigrant population, poor housing and relat-

---

[18] The Bayernhymne, Gott mit dir, du Land der Bayern, sometimes known colloquially as the Bayernlied, was written in the middle of the 19th century by Michael Oechsner (1816-93), music composed by Max Kunz (1812-75). The first verse is dedicated to the state of Bavaria and the second to the Bavarian people. The third verse, praising the Bavarian king, was dropped after 1918. From the 1950s onwards the song was taught and sung in Bavarian schools.

ed social problems, (certainly relevant to the 1989 Berlin situation) in times of economic difficulties.[19]

Forschungsgruppe Wahlen, the respected German electoral research institute, emphasises the protest vote nature of those who support radical parties like the Republicans, in wanting to give the established *Volksparteien* a warning, or something to think about (*einen Denkzettel verpassen*). Nevertheless, some experts think that this may be an under-estimation of the structural difficulties which German (and Bavarian) society experienced, particularly during the period shortly before and after German Unity. It is, therefore, probably still too early to say for sure whether a party like the Republikaner is going to become a permanent feature of either the Bavarian or the German party system. That did, however appear less likely after the poor performance of both the Republicans and all other right-wing extremist parties during "super election year" 1994.

After their success both in January, in Berlin, and in June, at the European elections, during 1989, the Republicans, along with other parties on the extreme right, suffered "a temporary eclipse", as Gordon Smith put it, whilst the Union parties benefitted from the brief euphoria surrounding the issue of re-unification. In the years since 1990, however, the Republican Party first recovered some of its former strength, for example Baden-Württemberg in 1992, by latching on to and exploiting some of the tensions created by German unity. By October 1994, however, electoral support had dwindled to less than two per cent nationally.

Other parties in Bavaria

A conservative ecological party, calling itself the *Ökologisch-Demokratische Partei*, (ÖDP) contested the 1986, 1990 and 1994 Bavarian state elections. The party tried to distinguish itself from the Greens by a more Christian-conservative approach to policies and programmes. It had little appeal.

The Bavarian party spectrum has included a wide variety of other, often very small regional groups from time to time. To name just three examples, the LIGA is a party which tries to project an image of itself as "a

---

[19] Gordon Smith. Dimensions of Change in the German Party System in Parties and Party Systems in the New Germany, ed. by Stephen Padgett. Dartmouth. Aldershot. 1993, pp.92/92.

party for life", campaigning mainly in Swabia and Upper Bavaria. The Augsburg Citizens Union (*Bürger-Union*) was formed by the former CSU deputy, Knipfer, and the Bamberg Citizens' Block (*Bürger-Block*) is another tiny party, operating only in Upper Franconia. There have, of course, been many such minor parties in Bavaria from time to time. Fifteen parties contested the last Bavarian state elections in September 1994, for instance – including the Natural Law Party (*Naturgesetzpartei*), which has appeared briefly in British elections. Most of these small parties achieve negligible success. A fairly recent German phenomenon, the "Instead" party (*STATTpartei – die Unabhängigen*) also appeared in Bavaria in the 1994 campaign, but to no avail.

The outstanding feature of the Bavarian party system nowadays remains the structural hegemony and political dominance of a single party – the CSU – which in the 1990 elections polled, not for the first time, more than twice as many votes as the SPD. Although the CSU now seems to have staged a political "come-back", after the double shock and initial loss of support associated with first the death of Strauß and then German Unity, the Bavarian polity was not always dominated by a "one-party system" to the extent to which it has been in the more recent past.

6

The Development of the
Bavarian Party System 1945-95

The new party system in post-war Bavaria evolved in an unusual way. Modern observers of Bavarian politics are sometimes surprised that the present dominance of the party system by one party was far from an immediate occurrence. The politics of Bavaria in the late Forties and Fifties were very different from those of later decades. With the benefit of hindsight, it is possible to identify different phases in the development of the Bavarian party system. These can be divided into approximately eight stages.

Stage I: May 1945-Dec. 1946

During a stage of nascence the formation of the first political parties took place under the strict control of the American Occupation Forces, who granted a *Land* licence to only five parties in the first instance, namely the CSU, SPD, WAV, KPD, and the FDP. This established a four-strand party system along the lines of a Communist/socialist/liberal/Christian-conservative model. Specific regional parties were not yet permitted. Bavaria's first minister president was Fritz Schäffer (CSU), appointed by the Americans, who later replaced him with Wilhelm Hoegner (SPD) on 28 September 1945.

Even at this early stage it appeared as if the CSU might dominate the proceedings, since it had 109 seats out of 180, following a vote of 58.3 per cent, in the Constitutional Assembly, elected on 30 June 1946. However the situation was complicated by the disagreements between the two factions in the CSU (see previous chapter).

Stage II: Dec. 1946-1950

When the people of Bavaria elected their first post-war parliament (*Landtag*) on 1 Dec. 1946, the CSU again polled an absolute majority (52.3 %). The seat distribution was as follows: CSU – 104, SPD – 54, WAV – 13, FDP – 9. Despite receiving 6.1 per cent, the KPD was unable to surmount the Bavarian ten per cent hurdle, which applied from 1946 until 1973, in any district. The author was told in several interviews in Bavaria that the original reason for the unique ten, rather than five per cent barrier was to bar the Communists. If that was the case, it certainly worked, because the KPD never got a foot-hold in the Bavarian party system and have played no part in it since 1946.

There were 180 seats in the first Munich parliament (*bayerischer Landtag*); every other Bavarian parliament since, right up to the present day, has consisted of 204 seats. One might have assumed that a two-party system, adopting clear Government (CSU) and Opposition (SPD) roles, would emerge. This was not the case. In the event the two largest parties formed a coalition with a third (smaller) party (see below). Unlike the party system soon to emerge in Bonn, the third party was not the liberal party.

In Bavaria the FDP was never able to play the pivotal role of "*Züinglein an der Waage*" (tipping the scales), as it was later to do at federal level. The four key functions of majority maker, liberal corrective, ideological balancer and transition agent in the federal party system, which enabled the tiny FDP to play a disproportionate role in Bonn were simply not available to the Liberals in Munich. In fact the Bavarian FDP was the only party to occupy the parliamentary opposition benches, in the first instance.

Because of the continuing ruptions in the newly founded CSU between the Catholic-conservative Hundhammer wing and the liberal-conservative Müller wing, it did not prove possible for the CSU to form a government on its own. Ironically it was the ultra-conservative, clerical wing of the party which proposed a CSU/SPD/WAV coalition,in order to prevent Josef Müller from becoming the first Bavarian minister president.

Müller could not muster the required support (50 per cent, plus 1) in his own party in an internal vote to select a candidate as minister president, even though he was party chairman. A compromise candidate, Hans Ehard, was chosen. Ehard went on to head Bavaria's first post-war coalition government.

In the first instance, this might have appeared superficially to be a cooperation stage in the party system, involving a Grand Coalition between the two major players on the Bavarian political stage, plus the WAV, a unique Bavarian phenomenon. In reality, it was more of an experimental or transitional stage – a desperate attempt to cope with the practicalities of an awkward political situation.

The first Bavarian cabinet, led by Ehard, consisted of four CSU ministers – finance, education (Hundhammer), agriculture and transport – four SPD ministers – home affairs, justice (Hoegner), economics and labour – and Alfred Loritz, the leader of the WAV, as minister without portfolio, and their secretaries of state. It did not include Müller.

Indignation and stormy battles soon developed. The SPD leadership (Schumacher and Ollenhauer) in the three Western zones objected strongly to Hundhammer's Catholic-conservative/*altbayerisch* education policy, and put pressure on the Bavarian SPD, who held leadership elections in May 1947. Waldemar von Knoeringen took over from Hoegner. An anti-socialist speech by Ehard in August was taken as the excuse needed to leave the coalition.

As the other partner, the WAV, had been abandoned on account of a political affair involving Loritz, Ehard formed a new cabinet on 20 September 1947 containing only CSU politicians. The ministerial changes included Josef Müller (Justice). Wilhelm Hoegner, whom he replaced, had actually been in favour of continuing the coalition. The CSU ruled alone until the end of the first four-year period of legislature. A stage of CSU dominance in the Bavarian party system seemed possible.

It was during this period that the CSU began to carve out its future dual role, as discussions in the Parliamentary Council in Bonn started in September 1948. Bavaria was represented by eight CSU delegates, four from the SPD and one liberal. Anton Pfeiffer from the CSU became chairman of the CDU/CSU *Fraktion*. Despite this joint cooperation there was some wrestling between Adenauer and the CSU over the federal structure of the envisaged German state. The Bavarians were insistent on promoting the maximum amount of federalism via a strong second chamber (Bundesrat). The CSU view prevailed.

Nevertheless the proposed Basic Law (*Grundgesetz*) was still not federalist enough for the CSU. In the decisive vote in the Parliamentary Council six of the eight CSU delegates voted against acceptance and two from Franconia voted in favour, along with the four SPD and one FDP delegates. In the subsequent votes in the Länder parliaments in 1949, Bavaria was the only state to reject the proposed constitution. In the Bavarian parliament 90 of the 101 CSU deputies voted against the *Grundgesetz*, 51 of the 53 SPD deputies voted for it, 8 of the 9 FDP deputies voted in favour and 7 of the 9 deputies from the Bavarian Party (licensed in 1948) voted against. The Bavarian parliament rejected the Basic Law by 105 to 66 votes (see below).

Table 6.1

Bavarian vote on the Basic Law 1949

	Total	For	Against	Abstentions
CSU	101	4	90	7
SPD	53	51	2	0
FDP	9	8	0	1
BP	9	1	7	1
Others	8	2	6	0
	180	66	105	9

There had been mounting pressure for Müller to be replaced as Chairman of the CSU. In May 1949 Hans Ehard, seen as "a man of the centre" who could best provide an acceptable balance between the political aspirations of the two warring factions in the CSU was elected as Party Chairman. After the vote against accepting the proposals for a new German constitution, the CSU was certainly viewed as a main player on the political stage in Bavaria by many Germans at the time.

Stage III: 1950-54

Some of the fundamental factors which affected the formative stage of the German party system, which Gordon Smith refers to as social and institutional factors[20], affected Bavaria too. As usual, however, Bavaria proved to be a special case. The WAV, a unique Bavarian phenomenon, disappeared completely in 1951, but the BP (*Bayernpartei*) played a brief role in the federal party system and a significant, more lasting one in Bavaria.

The Bavarian electoral law of 29 March 1949 increased the number of parliamentary deputies to 204 and gave each voter two votes. The BP, licensed by the Americans in 1948, immediately amassed great popularity amongst the Catholic farming and agricultural electorate in the rural areas of Old Bavaria. Josef Baumgartner left the CSU and proved to be a lively and popular leader of the Bavarian Party.

The BP polled 20.9 per cent at the 1949 federal election, giving it 17 seats in the first Bundestag, and making it the third strongest force in the Bavarian party system. This influenced the 1949 CSU federal vote in Bavaria (29.9 per cent). It is important to understand the crucial influence of the federal dimension on the Bavarian party system at a time when the CSU was just beginning to establish its vital dual role in German politics as both a federal party in Bonn and an autonomous regional force in Munich.

The election results on 26 November 1950 in Bavaria were also greatly influenced by the political debut of the BP, which firmly established itself as the third largest party with a vote of 17.9 per cent (39 seats). The appearance of the BP accounted for the loss of about half the CSU's membership in 1948. Most of the BP votes had been gained at the expense of the CSU, whose 1946 percentage of 52.3 was almost halved to 27.4.

This was the only post-war Bavarian election at which the SPD (with exactly 28 per cent) received more votes (around 60,000 more) than the CSU. Owing to two additional seats (*Überhangmandate*) gained by the CSU in Swabia in 1950 (the Bavarian electoral law of 11 August 1954 abolished additional seats), the CSU returned 64 and the SPD 63 deputies.

The BHE, a party founded first in Schleswig-Holstein and then in other

---

[20] Gordon Smith. Democracy in Western Germany. Heinemann. London. 1979, p. 103.

regions of Germany, represented refugees and expellees. It polled 12.3 per cent (26 seats) in Bavaria in 1950 and the FDP gained twelve mandates with a vote of 7.1 per cent (the 1950 WAV vote of 2.8 per cent marked the beginning of the end for the Economic Reconstruction League).

The Bavarian party constellation was clearly changing: a dramatic shift in emphasis occurred, as CSU dominance vanished at a stroke and the SPD technically became the largest party. A trend towards multi-partyism emerged, as new parties, which had not been granted a licence in the nascent stage, came on to the scene. There were several of these – even a monarchist *Königspartei* – but the astounding success of the Bavarian Party as a serious competitor to the CSU at both federal and regional level, was the biggest destabilising factor in re-shaping the party system.

Adenauer tried to persuade the CSU to form a "bourgeois" coalition in Munich on the federal model, but Ehard preferred the SPD and the BHE. The SPD leader von Knoeringen insisted that his party would not join a coalition government which included Hundhammer. He got his own way, despite considerable CSU protests. When Georg Stang died in May 1951, Hundhammer became president of the Bavarian parliament.

Ehard's third cabinet consisted of four CSU ministers (including Josef Müller and Hanns Seidel), three from the SPD (including Wilhelm Hoegner) and the BHE received two secretary of state posts. Friction occurred as the CSU were reminded periodically by the SPD, that the latter had gained more votes in Bavaria. Tension was heightened further by the results of the 1953 federal election (CSU: 47.9%, SPD: 23.3% of the Bavarian vote).

Adenauer continued to exert pressure on his Bavarian sister party to break away from the format of a Grand Coalition in Munich. He was determined to establish clear roles of Government and Opposition for the CDU/CSU and SPD respectively in the emerging federal party system. The Bavarian model did not fit into that pattern.

If the 1950 Bavarian state elections had brought changes in the party system, the events which followed the 1954 elections were to produce an even bigger surprise.

Stage IV: 1954-57

The Bayernpartei failed to clear the amended five per cent clause, applicable to the whole of West Germany instead of any one *Land*, at the 1953 federal election. It was therefore not allocated seats in Bonn on that occasion (or ever again). Although the BP's share of the vote in Bavaria declined at the 1954 elections as well (13.2 per cent – 28 seats), it was by no means a spent force.

The CSU increased its share of the vote to 38 per cent, giving it 83 mandates out of 204. The SPD repeated its 1950 performance, losing only two seats. The only other parties to clear the Bavarian ten per cent clause were the BHE (19 seats) and the FDP (13 seats); the Bavarian Liberals polled 13.2 per cent in their usual stronghold of Central Franconia, even though they could manage only 7.1 per cent in Bavaria as a whole.

The CSU simply assumed that it, as the largest party, would be able to dictate coalition terms. There was pressure from Bonn for the CSU to form a coalition with the BP, but the latter still remembered the fraternal feud in the late Forties, when it found itself fishing in the same pool of voters as the CSU in those parts of Catholic-conservative Bavaria which were more Bavarian than most.

During the election campaign the SPD had promised the electorate a new policy of Light across Bavaria (*Licht übers Land*), implying that the CSU's education policy, especially on denominational schooling and teacher training, belonged back in the Dark Ages. The FDP too felt very strongly about the CSU's "Christian" education policy (*eine christliche Kulturpolitik*), and pubished a paper entitled "katholisch, evangelisch oder christlich?" The CSU meanwhile sat back and waited. It was later accused by some in its own ranks of complacency and arrogance.

A unique Bavarian government was formed: the Coalition of Four (*Viererkoalition*) was the only government in post-war Bavaria not to contain the CSU. Wilhelm Hoegner (SPD) was elected minister president by 112 votes, as against 82 for Hanns Seidel (CSU). The party system was in a stage of disorientation and uncertainty, as the largest party was relegated to the opposition benches and Hundhammer, at the suggestion of the SPD, was replaced as President of the Bavarian parliament by Ehard.

A position of potential for alternation of the two biggest parties in the party system was realised with a political change in power (*Machtwechsel*). It should, however, be noted that, viewed with hindsight, this change in power proved to be very much a temporary phenomenon.

Four most unlikely bed-fellows, the SPD, BP, BHE and the FDP, fought long and hard, but in the final analysis unsuccessfully, against leading CSU politicians of the day (including Meixner, a prelate) and the Catholic Church – the 1924 Concordat between the Vatican and the Bavarian State was still valid.

The CDU/CSU parliamentary party in Bonn was not impressed by the polical situation in Bavaria. Matters came to a head when the Christian Democrats achieved an absolute majority – 50.2 per cent – in the 1957 federal election (15 September), at which the CSU share of the vote in Bavaria reached a staggering 57.2 per cent. Yet the CSU was in opposition in Munich. The Coalition of Four collapsed prematurely when the BP and the BHE each resigned from the government on 8 October 1957, after conducting "secret" negotiations with the CSU regarding a future Bavarian coalition following poor performances from both the small parties at federal level.

Stage V: 1957-62

Paradoxically it was precisely the CSU's bitter experience of its loss of power from 1954 until 1957 that heightened its resolve never to permit such a situation to recur. Viewed with hindsight the Coalition of Four can be labelled a watershed in the Bavarian party system. A new cabinet was quickly formed – this time there was less hesitation by the CSU! – on 16 October under Hanns Seidel, who led a CSU/BHE/FDP coalition which ruled until the end of the period of legislature.

The Bavarian FDP, who unlike the other two small coalition partners, did not abandon the SPD-led government in 1957, received its second and final experience as a party of government in Munich. In spite of any agreements which might have been reached, the BP, formerly the CSU's chief competitor for the votes of Bavaria's Catholic-conservative agricultural community, was *not* invited to join!

Following the state elections of 1958, the same coalition was continued. The CSU polled 45.6 per cent, compared with 38 per cent four years earlier. It received 101 seats in parliament – only two short of an absolute majority. The party was determined to consolidate its improved position and never allow the SPD to return to power in Munich again.

Although the CSU had persuaded the BP to abandon the sinking ship of the Coalition of Four in the autumn of 1957 with promises of possible cabinet posts in a new CSU-led government, it was equally resolved to see that the tiny BP reaped its just rewards for what the CSU leadership saw as the BP's treacherous behaviour vis-a-vis its "Catholic brother." The CSU saw itself as the only party in Bavaria which could lay claim to being "a Christian party."[21]

In the so-called Casino Affair (*Spielbankaffäre*) – a most unsavoury business – the BP politicians Geiselhöringer and Baumgartner were accused, originally in 1955, by the CSU of having profited personally through the granting of casino concessions. Allegations of bribery and corruption were made, with little or no hard evidence, prison sentences passed (but never served) and the reputation of the Bavarian Party was dragged through the mud until 1959/60. The party never recovered from that experience.

In interviews with Klaus Dehler[22] (the nephew of Thomas Dehler) and Hildegard Hamm-Brücher[23], both of whom were members of the Bavarian FDP at the time of the Coalition of Four, the author was told that the Casino Affair was one of the darker episodes of the period.

The period from 1958 until 1962 can best be described as an initial phase of consolidation of the CSU's position in the Bavarian party system. After the equilibrium of the party system had been disturbed between 1954 and 1957,

---

[21] "Da doch die CSU, die in Bayern der Alleinspruch darauf erhebt, als christliche Partei angesehen zu werden." Bremer Nachrichten, 4 November 1955, p.6.

[22] Interview with K. Dehler in Nuremberg on 9 March 1983. Dehler was the youngest member of the Bavarian FDP parliamentary party at the time (1954), aged 27. He was chairman of the Bavarian Liberals from 1964 until 1967, when he retired from politics.

[23] Dr Brücher, as she was in 1954, was a driving force in the FDP who fought hard for the establishment of the Viererkoalition. She married Erwin Hamm, a CSU politician, in 1956. Interviews with the author in the Bonn Foreign Office were on 25 May and 21 July 1982. Frau Hamm-Brücher was minister of state under Genscher, 1976-82, in the SPD/FDP government led by Helmut Schmidt.

the CSU launched a membership drive to compensate for a huge fall from over 82,000 in 1948 to some 35,000 in 1955. After Franz Josef Strauß took over sole responsibility for the *Bayernkurier*, the party's weekly mouth-piece, in 1961, a vast improvement in quality and circulation figures occurred.

The multi-partyism of the early Fifties was being replaced by a tendency towards party concentration, as the CSU increased its share of the vote at the expense of the small parties, and in 1958 the SPD recorded over 30 per cent for the first time. It was obvious that the party system in Bavaria differed radically from the one in Bonn. In this respect too Bavaria was an exception to the rule. In its early stages no balanced or consistent party system had evolved.

Stage VI: 1962-66

On 25 November 1962 the CSU again increased its share of the Bavarian vote (47.5 per cent), and achieved, for the first time since 1946, an absolute majority of the seats (108) in parliament. This, despite the Spiegel Affair in October 1962. Although F.J. Strauß was not forced to resign his federal government post as Defence Minister until 30 November, the imminent resignation was accompanied by much negative publicity immediately prior to the Bavarian elections.

The events in Bonn apparently did not affect events in Munich, where the popularity of Strauß, who had taken over as Bavarian party chairman twenty months earlier, seemed unassailable. The CSU's "institutional dual role" as Mintzel calls it, in the German party system was in evidence for all to see.

The Bavarian SPD also improved its position, gaining 35.3 per cent (79 seats) in 1962. The small parties were the big losers. The GP (*Gesamtdeutsche Partei*), as the BHE or DP-BHE was calling itself – the party had several versions of its name – failed to clear the electoral hurdle. The FDP (nine seats) and the BP (eight seats) both fulfilled the ten per cent requirement – though only just – in Central Franconia and Lower Bavaria respectively, as was their wont.

Initially there was some disagreement within the CSU concerning who should lead the next government, since Hans Ehard was stepping down because of his age. His deputy, Rudolf Eberhard, a Protestant from

Franconia, was interested, as was Alois Hundhammer, who was supported by the Catholic-conservative Petra circle. Strauß had also let his interest in the post be known. As a way out of the impasse, a compromise candidate, Alfons Goppel, was selected by the CSU as their candidate. Goppel won the vote in Parliament against Hoegner by 109 votes to 78 and became Bavaria's new minister president.

Ironically, Goppel the compromise candidate remained minister president until Strauß took over in 1978. Only Montgelas had held the reins of power in Bavaria longer at any time since the 19th century. Goppel was a jovial figure, who became increasingly popular and a real *Landesvater* to the Bavarian people.

This was a kind of transition stage in Bavarian politics – the transition between coalition governments and the time when the CSU finally established itself as *the* ruling party of state in Bavaria. Although the CSU had more than half the seats in parliament in 1962 and could have governed alone, for tactical reasons it officially formed a coalition government with the (now tiny) Bavarian Party. The BP had just scraped into parliament with only 4.8 per cent overall and 10.3 per cent in *Niederbayern*. However this was only a token coalition, since the BP had only one secretary of state post (Robert Wehgartner) in a cabinet of twenty-three.

On 20 July 1966, just before the end of the four-year period of legislature, Wehgartner resigned from the BP and joined the CSU. Ever since, right up to the present day, every Bavarian cabinet has contained only CSU politicians. The transitional phase in the party system was over; a new phase of state power monopoly by the CSU, the Bavarian "party of state and order" (*die bayerische Staats- und Ordnungspartei*) was about to begin.

Stage VII: 1966-88

This was a stage of complete dominance by the CSU of the Bavarian party system. Although it did not quite reach an absolute majority of the Bavarian votes (48.1%) in 1966, the CSU did again achieve over fifty per cent of the parliamentary seats (110 out of 204). However, the 1966-70 government under Goppel, was the first one – apart from 1947-50 – where one party ruled alone.

This phase marked the further demise of the small parties, except for the brief appearance from 1966 to 1970 of the NPD and the intermittent return to the Bavarian parliament, from 1970 onwards, of the FDP as a fairly impotent opposition party.

Of the two big parties, the CSU began to monoplise the party system to an extent which no German party has done in any area at any time since 1945. In 1970 the CSU, for the first time since 1946, won an absolute majority of both seats and votes has repeated the performance at every Bavarian state election right up to the present day. The party reached its zenith in 1974 when it polled an amazing 62.1 per cent. Two years later the CSU achieved exactly 60 per cent in the 1976 federal election.

It is easy to forget, given its more recent results, that the Bavarian SPD took over 30 per cent of the vote at every *Landtagswahl* between 1958 and 1982. In 1962 and 1966 it cleared 35 per cent. At these last two state elections it looked as though the former multi-party system in Bavaria with coalition governments was developing into a two-party system, with approximately twelve percentage points separating the CSU and the SPD.

That was not what happened. In 1970 the gap widened to 23 points, as the CSU went from strength to strength (56.4%) and the SPD lost some ground (33.3%). The overriding feature of the Bavarian party system in this stage of development, reinforced by the elections between 1970 and 1986 was CSU hegemony. The CSU achieved its aim of becoming the Bavarian party of state *par excellence*.

The West German party system between 1961 and 1983 was often described as a two-and-a-half party system, with the FDP as the half, in terms of influence, if not size. By analogy, the hegemony of the CSU in the Bavarian party system over the last three decades could justify the description one-and-a-half party system. It is, however, the SPD which is the half in Bavaria.

Such a description really would be no exaggeration. At the state elections in 1974 (62.1 as against 30.2) and in 1986 (55.8 as against 27.5) the percentage share of the CSU vote in Bavaria was more than double that of the SPD. At the intervening state elections the position was similar. In 1978 the percentage votes were: CSU – 59.1, SPD – 31.4, and in 1982: CSU – 58.3 and SPD – 31.9 per cent.

The years from 1969 until 1982/83 were a time of tension between Munich, where the CSU did not just rule, it "reigned supreme", and Bonn, where the SPD/FDP government had taken power. Since the Spiegel Affair relations between the CSU and the FDP were strained. Relations between the CDU and the FDP were better, and Kohl, when the CDU/CSU returned to power in 1982/83, deliberately used the Liberals as a way of restraining the influence of Straufl and his party at federal level.

Just one example of this will illustrate the point. Following the 1983 federal election, Kohl offered Strauß every ministerial post in his new government except the two which he knew the Bavarian leader would accept (Foreign Minister, Economics). There was considerable speculation at the time on whether FJS would return to politcs in Bonn. In a German television interview (ZDF) Strauß, in his usual ebullient style, said, in reply to the interviewer's question: *"Gehen Sie nach Bonn?"* (Are you going to Bonn?) *"Nein ich gehe nicht nach Bonn."* (No, I am not going to Bonn) and paused. The interviewer, taking the reply to mean that Strauß would not be returning to federal politics, was surprised at such a frank answer. The interviewee then added: *"Ich fliege nach Bonn."* (I am flying to Bonn), implying that he was not walking/going on foot (the literal meaning of the German expression), but flying – for discussions with Kohl the following day.

The dominant feature of this stage of the Bavarian party system was the hegemonic position of the CSU as an integrating factor in presenting itself as the sole party of state and government in Bavaria. The growing popularity of first Goppel, then Strauß, as the father of the Bavarian nation (*Landesvater*), underlined even further the unassailable position of the CSU and its minister presidents. The 70th birthday celebrations of FJS in 1985 rivalled those of a Bavarian monarch.

Stage VIII: 1988-93

The sudden and totally unexpected death of Franz Josef Strauß on 3 October 1988 brought the end of an amazing era in Bavarian politics. The era of party chairman Strauß (1961-88) had been one of unprecedented political development and unparalleled electoral success. The last few state elections in had taken on the aura of referenda on the personal popularity of the uncrowned "king" of Bavaria and the standing of Bavaria's (and Germany's) most successful post-war party.

As party chairman of the Bavarian CSU from 1961 until his death, Strauß's influence had spread through all echelons of the CSU hierarchy. He had been a dynamic driving force behind persuading the CSU to support the FDP/SPD request for a referendum regarding changes on denominational schooling in 1968 – the exaggerated influence of the Catholic Church in this area of policy had been one of the main reasons for the CSU losing power in 1954.

Strauß had striven to effect changes to the Bavarian constitution and the Concordat with Rome as part of a much-needed overall party modernisation strategy. The revival of the fortunes of the CSU in the late Fifties and early Sixties coincided with the rise of FJS, first as party chairman and later as both party chairman *and* Bavarian minister president from 1978 until his death.

The death of Strauß shook the Free State of Bavaria to its foundations. A huge power vacuum was created overnight. In the first instance Max Streibl, the deputy (*stellvertretender Minister Präsident*) took over, until a new cabinet could be formed sixteen days later. The initial reaction of most people was of course to ask who could possibly fill the great gap left by the departure of Bavaria's most popular and successful post-war politician.

In the period immediately following the death of Strauß many Bavarians were asking what future Bavaria could possibly have without its greatest leader (*Bayern ohne Strauß – was soll's?*). Others were hoping that the well-oiled technocratic party machine, which FJS himself had helped to modernise, could somehow carry on, with a new supremo at the wheel.

Max Streibl took over as minister president and Theo Waigel as party chairman (he was also federal Finance Minister). Streibl's first cabinet (19 Oct. 1988 – 30 Oct. 1990) soldiered on, but the CSU suffered two bad defeats in less than eighteen months of taking office. At the European elections in June 1989, less than nine months after Strauß's departure, the CSU vote fell from 57.2 per cent in 1984 to 45.4 per cent (1979: 62.5 per cent). The absence of FJS was apparently taking its toll (see chapter one).

The Bavarian local elections in March 1990 showed a drop of seven percentage points for the CSU, but also, perhaps more worryingly, the loss of control of seven county boroughs and several town halls to the SPD. The chances of "the new team" retaining the CSU's absolute majority at the approaching Bavarian state election in the autumn were being seriously questioned.

It was with tremendous relief that the CSU heard that it had polled 54.9 per cent at the Bavarian state elections on 14 October 1990. This was within one per cent of their result in 1986, when FJS was still at the helm. The Bavarian SPD was extremely disappointed with its 1990 electoral performance, which produced its worst ever result of 26 per cent.

Both the Greens (6.4per cent) and the FDP (5.2 per cent) entered the Bavarian parliament; yet it could hardly be called a four-party or even a two-party system with the CSU (127) having more than twice as many parliamentary seats as the SPD (58). The Greens were allocated twelve seats and the Liberals seven. All other parties, including the Republicans, were victims of the five per cent clause.

Despite the commendable CSU performance in retaining its hegemonic influence within Bavaria, doubts were being raised about the position of the Free State of Bavaria and the role of the CSU in the new, enlarged Germany. In the early Nineties there were also doubts expressed about whether Streibl was the right personality to lead the party into the "super election year" 1994.

When the "Amigo" affair broke – Max Streibl was accused of accepting free travel in private jets and limousines from and to various firms – the opportunity was taken to replace the man who was considered too much of a "backwoodsman," unknown outside Bavaria. Although he resisted to begin with, eventually Streibl had to bow to the wide-spread feeling that he was not the dynamic personality required to project Bavaria's image outside the Free State.

The Munich evening newspaper AZ (*Abendzeitung*) reported that Theo Waigel wanted the job of minister president, but that Stoiber would get it.[24] At the time (June 1993) opinion polls had estimated the level of CSU support to have dropped from around 55 to 43 per cent. Some thought that it was a mistake for Waigel, the party chairman to be a member of the federal cabinet,

---

[24] "Max Streibls Nachfolger – Waigel will, Stoiber wird's." Referred to in Spiegel 20/1993, p.19.

especially as it was claimed that he, the Finance Minister in Bonn, was associated in the eyes of sizeable parts of the electorate with the "tax lies" of Kohl's government and all the Republic's financial problems since re-unification.

There were further doubts about how Waigel and Stoiber would work together. Despite rumours that both men found the so-called "tandem" approach – the front man steering (the bicycle) and the other simply peddling – to joint Bavarian leadership uncomfortable (FJS had of course held both posts), they certainly presented a united front for the media and put on a first-class performance at the CSU party conference in Munich on 8/9 October 1993, at which a new party programme (*Grundsatzprogramm der CSU in Bayern*) – the first one since 1976 – was passed.

Waigel complained bitterly, and was apparently genuinely hurt by what he called a dirty campaign (*"eine Schmutzkampagne – eine Sauerrei"*) against his private life. As a "good" Catholic he had been living apart from his wife for a number of years and had a new partner. It was even alleged in the press that he had several illegitimate children.[25]

A stage of uncertainty regarding the CSU's way forward in the post-Strauß era remained during Streibl's second cabinet, which ended in June 1993. Another transitional stage in Bavarian politics came to an end.

Stage IX: 1993-95

The Stoiber era began in June 1993. Edmund Stoiber from Upper Bavaria has always been categorised as representing the clerical, ultra-conservative, far right of the party. Aged 51 when he took over the reins of power, Stoiber was considered by many Bavarians to be the sort of younger, more dynamic hardliner needed to lead the CSU into the bumper election year. The inner circle of the CSU obviously felt more confident of future success under Stoiber than Streibl.

Stoiber was dubbed "Edmund Thatcher" by some in late 1993/early 1994, owing to his anti-European stance. He expressed outspoken criticism of the Maastricht Treaty and his own government's policies on European integration[26]. However close observers of the occasionally stormy

---

[25] See Spiegel no. 21 1993, p.18ff

[26] See "Edmund Thatcher is no man of straw" by David Gow in The Guardian, 15 January 1994, p.4.

CDU/CSU relationship over the years, especially regarding the strategy followed by F.J. Strauß, will recognise certain common features in Stoiber's tactics. Stoiber has even described himself as the "alter ego of Strauß."

Like Strauß, Stoiber was not slow to let the CDU know in no uncertain terms that the CSU is still very much an independent party. In December 1993, within six months of taking office, Stoiber had accused Kohl and the CDU of doing far too little to integrate protest voters on the far right. He also claimed that the CDU was indistinguishable from the SPD on several issues. This was typical of the sort of political shadow boxing and tactical manoeuvering between the Union parties which occurred at regular intervals under Strauß.

This created an awkward atmosphere for Kohl and the CDU hierarchy when they visited the CSU headquarters in Munich in January 1994 for joint election strategy talks. The television pictures showed Kohl looking embarrassed when Stoiber presented him with the *Bayerischer Verdienstorden* (Bavarian order of merit). This is an award to recognise outstanding service to the Free State of Bavaria. Stoiber said that, since Kohl was born in the Palatinate at a time (1930) when it was still Bavarian territory, he was officially a Bavarian (!) and was receiving his award in recognition of his achievements regarding German Unity[27].

A new "negative record", as it was termed in the German media, was set with the announcement on 8 February 1994 of over four million unemployed for the first time in the Federal Republic of Germany. A disturbing trend was an increase in the number of engineers, chemists etc. with professional qualifications who joined the unemployed during 1993. This was followed, however, by a slight drop in unemployment in early 1995. The rate in Bavaria dropped to seven per cent in February 1995, remaining the lowest figure in any of the German federal states.

---

[27] A report of the meeting and presentation were shown on the Bavarian television news, Bayerische Rundschau, on 23 January 1994. The author had the distinct impression that Stoiber, whilst trying to lighten the difficult atmosphere between the two sister parties – given the Bavarian leader's controversial remarks on Europe just a few days earlier – was gaining the upper hand on his own territory. Kohl looked decidedly uncomfortable.

7

Bavaria's Booming Economy: From an Agricultural to a Modern Industrial State

In what in fact turned out to be his last public appearance, at the official opening of an aeroplane factory in Augsburg, Franz Josef Strauß, only three days before his death in October 1988, did not miss a golden opportunity – he never did – to boost the image of his beloved Bavaria. On that occasion Strauß said that he was pleased that Bavaria, formally an industrially and economically backward agricultural state had now become the centre of the German aerospace industry.[1]

As was shown in the opening chapter, Bavaria has throughout its history had a keen sense of being an independent nation state. Even today, as one of 16 *Länder* in the New Germany – albeit easily the largest territorially with an area of over 70,500 sq. km., and with the second largest population, now approaching 12 million, – it is worth reminding ourselves that the Free State of Bavaria is larger than many European countries. It occupies, for example, about the same territory as neighbouring Austria (population: 7.5 m.) and is almost twice the size of Switzerland (6.5 m.).

---

[1] This occasion was shown on the ARD German television news Tagesschau on 3.10.1988.

Bavaria in the modern European context

The Free State of Bavaria is larger territorially than five of the countries in the European Union – the Benelux countries, Ireland and Denmark. Bavaria also has a larger population than several member states of the EU: Denmark, Greece, Ireland, Luxembourg and the Netherlands.

Many Bavarians will tell you, quite correctly, that Bavaria is one of the oldest states in Europe. Even today they still compare Bavaria with other (sic) European countries, of which there are about forty-five. Bavaria is in twentieth position in terms of geographical size. In terms of population, it is thirteenth. With regard to volume of exports, Bavaria occupies twelfth position internationally, ahead of Sweden, Austria, Denmark and Spain (and even Australia).

As far as the gross domestic product per head of population is concerned, only Denmark is in front of Bavaria in the EU. Applying the same criteria, in 1990 Luxembourg was the only country in the European Community with a lower rate of unemployment[2]. Even as unemployment rose in the years following German unity, the rate of unemployment in Bavaria has remained, as usual, below the average for Germany, and well below that of almost all the other German federal states.

In 1992 Bavaria had the lowest unemployment rate (6.2 per cent) of any of the sixteen Länder, followed by Baden-Württemberg, with 6.5 per cent, at a time when the average percentage in the West was 8.3. The highest unemployment was in the five new Länder (average: 15.9 per cent), with the worst figure in Sachsen-Anhalt (17.6 per cent). In January 1994 unemployment in Germany passed the four million mark for the first time in the history of the Federal Republic – a new "negative" record, as the German media termed it. As usual, there was a huge discrepancy between the figures in the West (8.8 per cent), with just over 2.73 million unemployed, and the East (17 per cent), with slightly under 1.3 million out of work. The Bavarian figure was the worst for ten years, but still (though only just) below the west German average at 8.6 per cent.

At the beginning of 1995 Bavaria again had the lowest inflation and unemployment figures, followed closely by Baden-Württemberg, of any of

---

[2] Freistaat Bayern. Eine kleine politische Landeskunde. A brochure produced by the Bayerische Landeszentrale für politische Bildungsarbeit. 1992.

the sixteen *Länder* in the new Germany. In January 1995 the lowest German unemployment figures were 7.2, 7.5 and 8.2 per cent in Bavaria, Baden Württemberg and Hesse respectively and the highest were 16.5, 17.0 and 17.6 per cent in Thuringia. Mecklenburg, West Pomerania and Saxony-Anhalt respectively.

Diversification

Bavaria's transformation from a predominantly agricultural state after the war to a modern industrial economy with an important service sector was an amazing example of economic diversification. Dr Tyll Neckar, a former president of the Confederation of German Industry (*Bund der deutschen Industrie*) said of the Bavarian economy:

"Whilst Bavaria developed from an agricultural state into a high-tech state, other regions in the Federal Republic were still in the grips of structural crises which were hard to overcome . . . all in all Bavaria is in this way an impressive example of successful structural change. Instead of retaining old structures, new ones are promoted."[3]

In 1882 over half the Bavarian workforce was employed in agriculture. After the second world war about one third of all employees worked in the primary sector (agriculture and forestry) in the Free State of Bavaria. Nowadays the figure is around six per cent, with agriculture and forestry contributing only around 2.5 per cent to Bavaria's gross domestic product.

The share of Bavaria's total working population in 1950 (4.2 million approx.) in agriculture/farming and forestry – the primary sector – was 34 per cent. By 1970 it had sunk to 13.2 per cent, and by the time the new Germany was formed in 1990 it had dropped to 6.3 per cent.[4]

This does not, however, mean that agriculture is no longer a significant factor in the economy. On the contrary, Bavaria, which possesses nearly 29 per cent of the land available for agriculture in the old *Länder*, changed from being a state of agricultural imports to one of exports, and technical progress and vastly increased efficiency mean that nowadays more can be

[3] Information über Bayern. Brochure published by the Bavarian State Chancellory. 1991. p.46.

[4] Rainer A. Roth. Freistaat Bayern. Politische Landeskunde. publ. by the Bayerisch Landeszentrale für politische Bildungsarbeit. München. 1992, p.191.

produced, even with a reduced workforce in the primary sector, i.e. agriculture/farming and forestry. This area of the economy still employs more people in Bavaria than the average for the old federal states.

Bavaria's economic miracle

As the number employed in the primary sector decreased, so there was a corresponding expansion in the number of Bavarian workers in manufacturing industry, the secondary sector of the economy, and the service industries (the tertiary sector). The latter includes areas such as banking, insurance, tourism, transport, administration and social services.

In 1950 roughly one third of the Bavarian work-force was employed in each of the three main sectors. By 1970 the manufacturing trades accounted for just over 45 per cent of those in employment in Bavaria, with just under 40 per cent in the services sector. In 1990 the share of the Bavarian work-force in the primary sector was even lower – only around six per cent, as we have seen – employment in the manufacturing sector dropped only slightly, but the service industries' share in the late 1980s increased to over 50 per cent (1990: 51.6 per cent). The five per cent drop in manufacturing (1970: 47.2 per cent; 1990: 42.1 per cent) was caused by the introduction of new processes involving greater automation and rationalisation.

Changing work patterns amongst women

The structural changes in Bavarian society due to economic factors have affected the work patterns of women particularly, although there has been a marked reduction amongst both men *and* women in the number of self-employed workers generally. Fewer women in Bavaria nowadays "help out" in family concerns, but whilst the numbers of female manual workers (*Arbeiterinnen*) has not changed drastically, a dramatic increase **has** occurred in the number of women who work as salaried employees (*Angestellte*) and civil servants (*Beamtinnen*).

The changes which occurred in Bavaria between 1950 and1990 with relation to the three main sectors of the German economy affected female workers to an even greater extent than males. In 1950 the percentages of women working in the primary, secondary and tertiary sectors of the

Bavarian economy respectively were 44.8, 23.4 and 31.8. By 1989 the respective figures were 7, 30 and 62,3.[5] In the case of married women, in 1950 around one in three worked outside the home, but by 1987 it was one in two. Here again structural changes in Bavarian society combined with those in Bavaria's economy.

From strength to strength

In Bavaria today just over 40 per cent of its GDP comes from manufacturing trade and more than 55 per cent from the service or tertiary sector. Tremendous diversification and an open attitude to new branches of industry have been the hallmarks of Bavarian success in this respect. Despite the revolutionary changes undergone by the structure of the economy, Bavaria remains the largest food producer in the Federal Republic.

In the immediate post-war period, however, Bavaria's economic outlook was indeed bleak. Bavaria had been separated from its traditional trading markets in the east and to the south by the iron curtain. It lay "tucked away" in the south-east corner of the free market of western Europe, rather cut off. Bavaria had no coal, steel of other raw materials to speak of; its only chance was to develop "raw materials of the mind," by means of promoting modern scientific knowledge and new technology, supported by research.

During the first two decades of the Federal Republic (mark one) a North-South divide (*Nord-Süd Gefälle*) developed. In the northern, export-oriented ports of Hamburg and Bremen, and in the Ruhr area, centred around Duisburg, Europe's largest inland harbour, the old industries of ship-building and coal-mining/iron and steel, respectively, flourished. The south, by contrast, was at that time associated with peasant farmers working the land, agricultural produce and a generally poorer economic outlook.

During the second half of the former FRG's existence, however, the roles were reversed, as many shipyards and steel plants closed and the northern cities and the Ruhr were badly hit by unemployment; it became more accurate to speak of a South-North divide, as a completely new type of industrial base began to develop in the southern half of Germany.

---

[5] ibid,. p.192.

In 1985 this was referred to in an article in a British newspaper (the Times) as the difference between the "sunbelt south and rust belt north." The same article, with some justification, despite the over-simplification, mentioned "glossy new aerospace and robotics plants nestling in pretty countryside" in the country's most southern states.

Although areas like Hamburg and the *Ruhrgebiet* have now developed new industries to replace the old ones, life-style and employment patterns were seriously affected at the time. Bavaria today, despite some minor set-backs, remains a very attractive location for the establishment of new businesses. As such, it is now one of Europe's most dynamic economic regions, attracting considerable investment.

In 1994 Hamburg was confirmed as the wealthiest city in Europe. In a comparison of greatest divergence between the rich and the poor in the European Union, Germany was the country with the largest gap between prosperity in Hamburg and poverty in Thuringia.

In the mid-eighties unemployment in the five northern *Länder* of the FRG was almost 50 per cent higher than in the southern ones. Economic growth in Bavaria, as well as Baden-Württemberg, during 1970-1986 was around 40 per cent greater than in federal states like North Rhine Westphalia, Hamburg and Bremen. New micro-chip and computer-related industries sprung up at an incredible rate in the new "silicon valleys" of South Germany. As a result, many unemployed Germans moved south to find jobs in what for many was a more attractive environment, both in terms of life-style and scenery, as well as employment opportunities.

Foreign companies such as Hitachi, Texas Instruments, Motorola and many others invested in Bavaria, as well German giants like Siemens, who had moved their headquarters from Berlin to Munich after the war. At the time of reunification, Siemens employed around 100,000 people in 25 factories in Bavaria alone, which represented about one third of their total staff world-wide, although more recently some workers have been laid off.

In an examination of changes in industrial density based on figures from the Bavarian Office of Statistics, Mintzel[7] demonstrates that between 1952 and 1973 industrial density in the Federal Republic increased by 14

---

[7] Alf Mintzel. Die Geschiche der CSU. Ein Überblick. Westdeutscher Verlag.Opladen. 1977, p.35.

per cent and in Bavaria by 45 per cent. There were quite staggering increases in each of the seven Bavarian districts, ranging from 25.9 per cent in Upper Franconia to 120 per cent in Lower Bavaria. The latter had traditionally been the most backward of the seven districts in economic terms and was therefore the district which made the most rapid progress when the economy took off.

The economic miracle in Bavaria was unique. It not only skilfully streamlined and modernised the established Bavarian economy, such as it was, it also developed completely new industries with an eye to the future. Between 1950 and 1979 Bavaria's gross domestic product increased by over fourteen times.[8]

Bavarian arts and crafts

Arts and crafts (*Handwerk*) have always occupied a special place in the Bavarian economy. In 1949 there were some 9,000 small businesses associated with arts and craft trades in Bavaria. Some of them had existed for centuries: wood and ivory carvers, violin makers, organ-builders, craftsmen working with pewter and wrought iron, locksmiths, fitters, glassblowers, traditional costume tailors, hand weavers and embroiderers, basket makers, boat-builders and many more. Some of these craftsmen can still be found in Bavaria today, but in much smaller numbers of course.

During the early years of the Federal Republic the Bavarian craft industries were re-organised into 125 branches, according to profession, distributed across the established craft groups, e.g. construction, metal, wood, clothing, textiles and leather, food, health, glass, paper, ceramics etc. As well as changes within and between the various craft trades, a small service industry of its own grew up, as mechanics and servicing engineers were required for installation, assembly and after sales service in many fields.

With the new industries in postwar Bavaria, a whole new area of trade developed for sub-contractors and suppliers. In addition, the growing demand for large quantities of consumer goods meant that the old crafts had to be adapted to, at the very least, more modern manufacturing methods and, in some cases, mass production.

---

[8] Helmut Hoffmann. Bayern. Handbuch zur staatspolitischen Landeskunde. Olzog. München. 1981, p.210.

Many people in Bavaria who had always worked with their hands found themselves working in another trade, perhaps now servicing vehicles, central heating or insulation systems. The former village blacksmith might have found himself responsible for the maintenance and repair of a large amount of agricultural machinery, as farming methods were modernised.

Craft and artisan trades, such as glass-blowing, porcelain, ceramics and the like were subject to expansion and modern marketing techniques. Although some craft trades were no longer in demand, with growing prosperity in post-war Gemany came a demand for quality and precision instruments and a wider variety of specialised products, sometimes individually commisioned. Increasing tourism brought the demand for an increased number of people working in jobs such as photographers, hairdressers, goldsmiths, turners, furriers and specialist tailors.

The EEC, as it was then, supported the concept of the arts and crafts being recognised as an indispensable constituent part of a country's economy. Fresh needs always arise, and it is a basic principle that the consumer should have the choice between cheaper mass-produced goods and more expensive, individually styled products from a trained craftsman.

This very important branch of the Bavarian economy, which is responsible for giving Bavaria part of its special traditional character, has made a valuable contribution over many years and continues to do so.

Medium-sized firms predominate

An examination of Bavarian industry some forty-five years ago would have revealed a tremendous number of small businesses, many of them one-man outfits of the type referred to above. Even in the period between the two world wars, the Bavarian economy had been mainly dependent on agriculture and small and medium-sized businesses, with far fewer large companies than some other areas in Germany.

After 1945 Bavaria continued its long tradition of encouraging personal initiative and enterprise via small and medium-sized businesses (defined as having less than five hundred employees). In 1961 the Bavarian small business loans programme was introduced. Its grants and loans triggered a high volume of investment over many years. In 1974 Bavaria was the first region in Europe to pass an act promoting the feasibility and productivity of small and medium-sized businesses as the driving force of the social market economy.

In 1988 the Bavarian Economics Ministry published a report which stated that approximately 98 per cent of the total number of businesses in Bavaria at that time fell into this category. Around 55 per cent of all employees in Bavaria were working in such firms (*mittelständische Betriebe*), which accounted for over 40 per cent of the total turnover the economy in the trade and commerce sector of the Bavarian economy. About 85 per cent of Bavaria's trainees and apprentices in 1988 were working in that area, and more recent surveys confirm that the situation in Bavaria has not changed significantly.

Small and medium-sized businesses in Bavaria have, on the whole, adapted well to greater specialisation; they have also made a strong contribution to the Bavarian export drive. The bankruptcy rate of businesses in this category is usually fairly high; in Bavaria, however, it is below the national average for Germany.

It would be difficult to over-emphasise the importance of this sector of the Bavarian economy. Since this type of work frequently offers employment close to home, it is advantageous to women. This has clearly been an important factor in influencing rural life, which has indeed been better preserved in Bavaria than in some other German states, where, in some cases, almost all the younger inhabitants have deserted the villages in search of jobs in the city. Many feel that this has helped to preserve some of the attractive features of community life as a counter balance to some of the less attractive aspects of modern life in the big cities.

Small and medium-sized businesses frequenty offer flexibility and an advantageous working atmosphere. Because capital outlay and return are crucial factors, a salient feature of Bavarian economic policy (*eine gezielte Mittelstandpolitik*) over the years has been the specific targeting of this essential area of the economy, in order to further its dynamism, productivity and competitive edge.

This resolute refusal in Bavaria to concentrate only on the big names in business – preferring instead to retain a human face – is seen by some as part of an overall concern for the individual and and the working environment of the majority. A recognition of the vital role played by small and medium-sized firms in the Bavarian economy is linked to a determination not to abandon the values and abilities of the skilled craftsman and the small businessman. Nevertheless, as we have seen, one or two economic giants with international reputations exist too. In Bavaria the two are not regarded as mutually exclusive; they function smoothly, side by side.

The political dimension of economic policy

Bavaria's economic miracle, within the context of what was happening in the Federal Republic generally, was naturally linked to political development. As Mintzel points out, the rise of the CSU as the ruling party of state was inextricably connected to a modernisation process in Bavaria[9]. Mintzel speaks of the consolidation phase from 1949 to 1957 when the CSU fought a fraternal feud (*Bruderzwist*) with its rival, the Bavarian Party. The BP was the representative of a strongly Catholic, Old Bavarian, agricultural tradition, and as such was a serious competitor to the CSU, which was determined to become **the** Bavarian party of state *par excellence*.

On its way to becoming Bavaria's major(ity) state party (*Staats- und Mehrheitspartei*), the CSU cleverly aimed for a compromise solution, trying to steer a middle way between, on the one hand, being faithful to Bavaria's enduring traditions and its pride in past achievements, and, on the other hand, attempting to bring about the much-needed progressive approach and modernisation programme.

Since Bavaria in 1945/46 was a predominantly agricultural *Land* with one or two "islands" of industry – Munich, Augsburg, Nuremberg, Fürth, Schweinfurt – the Bavarian economy, under the CSU's leadership, developed gradually into a modern high-tech industrial state with reserves of agriculture. Structural changes in Bavarian society accompanied these economic developments.

In political terms, the main agent for change was the CSU, which during the 1950s absorbed its main Bavarian competitor, the *Bayernpartei*. The BP failed to adapt to changing circumstances; instead it chose to cling obstinately to the outdated structures associated with the purely farming and

---

[9] Alf Mintzel. Die CSU in Bayern als Forschungsobjekt in Oskar Niedermayer und Richard Stöss (Hrsg.). Stand und Perspektiven der Parteienforschung in Deutschland. Westdeutscher Verlag. Opladen. 1993, p.90 ff.

agricultural areas of predominantly Catholic Old Bavaria.

Meanwhile the CSU had gradually begun to dominate Bavarian politics. By the early Seventies, when the ruling party had an absolute majority of both votes and parliamentary seats in Munich, only small sections of Bavaria, along with the whole of Scotland, Wales, Ireland, southern Italy and western France, belonged to the poor regions of the European Community.

Agriculture and forestry today

Between 1950 and 1990 the number of agricultural holdings in Bavaria was reduced from 440,300 to 224,000[10]. Nevertheless, over the same period of time, the amount of land cultivated for agricultural purposes remained approximately the same, which meant that the average size of an agricultural holding almost doubled. In order to compensate for the reduction in the number of people in Bavaria working in the primary sector, great effort was put into introducing more mechanisation and improving farming methods generally. That involved, for example, varying cultivation and harvesting techniques and the use of fertilisers etc.

As a result, *fewer* agricultural holdings employing *fewer* workers overall produced *greater* yields. Bavarian farmers produced record achievements. In the case of sugar, beef and cheese, Bavaria now meets all of its own requirements, and Bavarian farmers provide around a third of milk and beef, a quarter of wheat and nearly a fifth of pork requirements for west Germany. About 95 per cent of German hops and one quarter on the international market come from Bavaria.

Between 1973 and 1983 exports of Bavarian produce generally tripled, especially to Italy, France and the USA. Exporting of foodstuffs increased dramatically and nowadays Bavaria exports agricultural goods worth over seven billion DM to more than thirty countries.[11]

The main agricultural produce of Bavaria's seven districts is as follows: Lower Franconia – wine-growing, especially *Frankenwein* in the Würzburg region, sugar beet, fruit, wheat, horticulture. Upper Franconia – fish, fruit, horticulture and beef, wheat. Central Franconia – hops, pork, fish. Upper

---

[10] op.cit. Roth, p.194.

[11] Figures taken from various sets of statistics provided by the Staatsministerium für Ernährung, Landwirtschaft und Forsten.

Palatinate – beef, fish, milk, potatoes, sugar beet, wheat. Lower Bavaria – pork, hops, sugar beet, beef, wheat, horticulture. Upper Bavaria – potatoes, hops, wheat, beef, milk, horticulture. Swabia – pork, beef, milk, horticulture. Rape and rape oil have become more important recently.

An increased contribution from federal funds and advantageous subsidies from the European Community's Common Agricultural Policy helped to counteract rising costs in the time before German Unity. In the early Nineties, however, Germany faced astronomical financial problems in trying to re-generate its eastern economy. The agricultural holdings in the former GDR, which were state-run as large co-operatives, presented a special problem.

Forestry as an economic factor

Approximately one third of Bavaria's total area is covered with forest, two thirds of Bavaria's farmers own woodland, and Bavaria has more forest and woodland than any other German federal state. About half of it is now privately owned, a third belongs to the Free State of Bavaria and the remainder belongs either to corporate bodies or to the German federation (*Bund*). The wood industry and areas of the economy associated with it, such as sawmills, processing and furniture production, are an important aspect of Bavaria's export and tourist trade nowadays. Many of the retail outlets in the more commercialised tourist areas are overflowing with Bavarian speciality products made from wood.

When the national park of the Bavarian Forest was opened in 1970, it was the first one in the Federal Republic. Bavaria also has the Berchtesgaden Alpine national park. These two national parks and the seventeen nature parks, together with over 400 nature reserves and more than 700 other protected areas, occupy nearly a quarter of Bavaria's total territory. Against a background of dying forests (*Waldsterben*) in the Seventies, Bavaria passed a forestry law in 1975, updated in 1982. In 1984 the protection of nature (water, air, soil, forests, animals, plant life), monuments and free access to areas of natural beauty were anchored in the Bavarian constitution, art.141.

In Bavaria tremendous importance is attached to formulating comprehensive policies and detailed regulations, even at local level, concerning the protection of the environment. In 1970 the Bavarian government was the first one in Germany to establish a special ministry for country planning and environmental questions (*Staatsministerium für Landesentwicklung und Umweltfragen*). This ministry, first headed by Max Streibl, has existed ever since. Various aspects of the great concern in Bavaria for the environment have had an impact on another key area of the economy – tourism and communications.

The tourism industry

Bavaria is Germany's most popular holiday destination for both Germans and foreigners, followed by Schleswig-Holstein and Baden-Württemberg. Bavaria's beautiful scenery, summer and winter sports facilities, cultural riches and numerous spa towns and health resorts have attracted growing numbers of visitors over the years.

Every fifth visitor to Bavaria now comes from abroad; the statistics reveal particularly large numbers of visitors from the USA, the Netherlands, Italy, Austria and Japan. The Bavarian tourist industry takes its customers very seriously, since travel and tourism nowadays provide over 300,000 jobs in hotel, catering and related branches in Bavaria.

Despite the general boom in this industry, in 1993 a ten per cent reduction in the number of visitors to Bavaria from abroad was recorded. As every fifth tourist is in the over-sixty age bracket, owing to a reduction in the retirement age in many countries, the Bavarian Tourist Board announced in 1994 that it would be targeting this age group particularly in future.

An excellent infra-structure, including a first class communications network, underpins the Bavarian tourist industry. Since the completion in 1992 of the final link in the Rhine-Main-Danube canal, which runs through Bavaria and connects the North Sea with the Black Sea, enormous development opportunities are available. The new canal has emphasised the ideal position in which Bavaria now finds itself, able to act as a vital player on the stage of promoting trade links and developing contacts between East and West in the new Europe.

Nearly fifty very well equipped health resorts and spas, with medicinal baths and excellent recuperation and rehabilitation facilities, account for around a third of overnight bookings in Bavarian hotels and guesthouses. Despite recent changes and cut-backs, the German health service still offers financial support to many patients whose doctors have recommended treatment (Kur) in such resorts.

Bavaria nowadays spends huge amounts of money on marketing its attractive image both at home and abroad. It has so far, even in the face of stiff competition, continued to attract vast numbers who visit Bavaria not only on holiday but also on business, often attending one of the growing number of trade fairs and exhibitions.

Special features of the flourishing Bavarian economy

As Bavaria became part of the new Germany in 1990, economic performance reached a new record level of 5.4 per cent growth in real terms, compared with 4.7 per cent in the FRG overall. From 1982 to 1991 year-on-year economic growth in real terms in Bavaria was steady, never in the red, and always slightly ahead of the figure for the Federal Republic as a whole. From 1970 to 1990 the rate of economic growth in real terms amounted to 87 per cent – the highest amongst the west German federal states (average: 61.4 per cent).

In manufacturing industry a similar situation obtained. Both Bavaria and the FRG overall experienced minus growth in 1982, but from 1983 to 1991 there was positive economic growth every year, oscillating in Bavaria between 1.1 per cent in 1987 and seven per cent in 1985 and 1989.[12]

Manufacturing was the sector which expanded most rapidly at the time of the economic miracle (*Wirtschaftswunder*), and over the period 1950 to1970 it provided the greatest number of jobs in Bavaria. Not just "made in Germany" but "made in Bavaria" also became a trademark synonymous with high quality products. The range of products is enormous. Suffice it to mention only a few of the main ones: office equipment, electronic apparatus for data processing, vehicles such as cars (BMW in Munich and Regensburg, Audi in Ingolstadt), lorries and buses (MAN in Nuremberg), heavy-duty machinery, a wide range of high-quality consumer goods, e.g. porcelain, china, ceramics and glass from Upper Franconia and the Upper Palatinate.

---

[12] All the figures quoted here, and in the parargraph above are taken from tables supplied by the Bavarian Economics Ministry (Staatsministerium für Wirtschaft und Verkehr).

Structural economic policy was also targeted at particular regions. The Upper Palatinate and Lower Bavaria were traditionally the least developed regions economically. For some years they were neglected, as was regional development policy within Bavaria generally. Between 1970 and 1986, however, the gross domestic product per head of population in the Upper Palatinate and Lower Bavaria increased at an amazing rate, in fact more rapidly – by 19.7 and 27.7 per cent respectively – than in any other of the seven regions. The percentage increase over this period for the whole of Bavaria (11.2) was considerably higher than for the Federal Republic as a whole (2.2).[13]

A high demand for Bavarian products abroad meant that during the Eighties exports rose from 28 to over 32 per cent, again slightly ahead of the very impressive figure for German exports as a whole. Around seventy per cent of Bavarian exports were either from the vehicles and machinery branch, or from electrical engineering (Siemens in the Erlangen/Nuremberg region and Munich) and the chemical industry.

The tremendous expansion in the micro chip, electronics and electrotechnical industries and the numerous computer-related branches has now become almost a Bavarian – certainly a southern German – speciality. Bavaria has also become a centre for the German aerospace industry, as the late Franz Josef Strauß, himself an amateur pilot, often pointed out. Munich's new international airport, which opened in 1992, is named after him.

In 1969 three Germans (the aircraft designer Willy Messerschmidt, based in Bamberg, the entrepreneur Ludwig Bölkow, based in Ottobrunn, south-east of Munich, and the shipyard founder, Hermann Blohm, based in Hamburg) joined forces and formed Messerschmidt-Bölkow-Blohm (MBB), Germany's biggest aerospace and armaments company. The ownership structure was originally extremely complicated. MBB used to be owned by the *Länder* of Bavaria, Hamburg and Bremen, as well as a number of companies, including Siemens and the Bayerische Vereinsbank.[14]

Large companies, of course, often have a very wide range of products and operate in more than one field. MBB, for example, was involved not only in the production of the European Airbus – although that work was not carried out in Bavaria – but also in space satellites, tank defence systems,

---

[13] See op.cit. Roth, p.200

[14] Information taken from das Hamburger Abendblatt, 9.09.1989.

multi-purpose helicopters, commercial aircraft. The company even produces the robust materials used for the futuristic magnetic levitation monorail train, which can travel at up to 220 miles per hour. MBB also experiments with laser technology for hospitals, pioneered by Dornier, another Bavarian company. Dornier also makes advanced medical equipment – a special type of wave bath which offers patients a new treatment for kidney stones, rendering the standard operation for their removal unnecessary.

Research and development

Another reason for Bavaria's remarkable success in keeping abreast of so many new developments is the importance attached to research. For many years the Max-Planck-Institutes in Garching, and thirteen other locations around Bavaria, as well as the Fraunhofer-Gesellschaft for applied research, with eight different institutions plus other research institutes, have been innovative in many fields. These include, amongst others, laser technology, satellites and telecommunications, medicine, plasma, quantum physics and optics, biochemistry, genetic engineering and radiology.

The different research institutes regularly team up with one or more of the eleven universities in Bavaria, where various research departments link up with local companies, in order to help develop new products for industry and advise on their practical applications. Bavaria, as well as Baden-Württemberg, has an especially strong reputation throughout Europe for the development of high technology in the field of microelectronics. This research is located in the Munich-Nuremberg-Stuttgart triangle.

Energy provision

Bavaria has few natural resources and is situated a long way from any coal mines or coastal ports. These disadvantages in the past were overcome via strenuous efforts by successive Bavarian governments to pursue a modern energy policy.

Originally water from Bavaria's lakes and rivers was the region's most significant source of energy. The building of Germany's first big storage power station at Lake Walchen and the hydro-electric plant on the river Isar were pioneering achievements for Bavaria. In the 1960s Otto Schedl, who was economics minister at the time, succeeded in promoting the development of an oil economy, based in Ingolstadt, with pipelines stretching to the big Mediterranean ports.

In 1966 the first commercial nuclear power station in the FRG was started in Grundremmingen, Bavaria. After the oil crisis of 1973, Bavaria relied less on oil as an energy source. Between 1950 and 1990 coal and lignite/brown coal (one of Bavaria's few natural resouces) became much less important, dropping from 37.7 and 24 per cent respectively to only 5.4 and 2.7 per cent of the primary energy sources exploited in Bavaria.[15]

Although alternative energy sources are being explored – for example the state of Bavaria is investing DM 30 million in the solar-hydrogen pilot scheme in Neunburg – practical application has proved limited so far. After hydro-electric power was extended to its limits, gas and nuclear power were relied on more heavily in the Seventies and Eighties.

In 1992 nuclear energy was responsible for around 65 per cent of the Bavarian electricity supply. The Bavarian authorities regard nuclear energy as ecologically sound, since the carbon dioxide content has been drastically reduced. KWU in Erlangen, a Siemens subsidiary, is considered able to produce a safe and environmentally friendly product. Bavaria now boasts the lowest electricity prices in Germany.

There is currently a campaign to raise public awareness, both in Bavarian domestic and industrial consumers, of effective energy savings and the importance of a forward-looking energy policy. It is, as always in Bavaria, linked to a policy which lays great stress on protection of the environment.

An economy of great contrasts

Just as many aspects of Bavarian life are full of apparent contradictions and contrasts, the Bavarian economy is no exception. Although traditional craft industries such as the production of toys, jewellery, china, porcelain, ceramics and glass still thrive, Bavaria is now the federal state with the greatest density of banks. Apart from Frankfurt am Main, the German business and banking capital, Munich now has more banks than any other German city, including the biggest Bavarian bank, *die Bayerische Landesbank Girozentrale*. This is the second biggest of the Federal Republic's *Landesbanken*. Three of the seven largest banking concerns in Germany now

---

[15] op.cit. Roth, p.211.

have their headquarters in Munich. Two of the oldest Bavarian banks are the *Bayerische Hypotheken- und Wechselbank*, established in 1835, and the *Bayerische Vereinsbank*, founded in 1869. Both operate as commercial banks, as well as specialising in mortgages and property loans.

The rapid expansion of the service sector made the Bavarian capital, with over 1.3 million inhabitants, the number one city in the insurance business in Germany (ahead of Cologne and Hamburg) and number three in Europe (after Paris and London). Munich's leading position is now clearly in evidence from its ultra-modern skyline. Some of these gleaming "palaces" house the headquarters of the top Bavarian and German banks and insurance companies.

Since 1935, when the two stock exchanges in Augsburg, originally the major one, and Munich merged, the Bavarian capital has been the location for the Bavarian Stock Exchange, which, before reunification, ranked third to Frankfurt and Düsseldorf. Munich today is viewed by many as the financial metropolis of southern Germany. There could be no starker contrast in the physical environment of these palatial edifices, representing the fields of insurance and banking, than with a quiet farmyard or a village church nestling in the peaceful Bavarian countryside. The urban-rural split is particularly apparent in Bavaria, even today. Indeed, it is felt by some visitors to be part of the region's special charm.

There is also a sharp contrast between the craftsmanship still employed in the manufacture of some Bavarian products, for example children's toys, and the modern marketing and sales techniques in evidence at today's trade fairs and exhibitions. Both Munich and Nuremberg have an international reputation in this respect. At the 45th international Toy Trade Fair in Nuremberg in 1994 competition was so fierce that 800 firms who wanted to display their wares were unable to do so.

The staggering transformation in the Bavarian economy from a predominantly agricultural state to one of the most modern industrial regions in Europe has, without doubt, been a major factor in boosting Bavaria's reputation both at home and abroad. With the introduction of the Single Market in 1993, Bavaria was ideally placed, not just in the new Germany, but in the centre of the new, more open European market place, to take advantage of expanding trade links between East and West.

8

A Study of Electoral Behaviour in Bavaria in 1990

The first detailed and comprehensive survey of electoral behaviour in Bavaria after German Unity was carried out shortly after the Bavarian state elections, held on 14 October 1990[1]. The data was gathered during October and November 1990. The survey was conducted by the Mannheim opinion poll research institute Forschungsgruppe Wahlen and compiled under the auspices of the Hanns-Seidel-Stiftung (foundation) in Munich. The analysis of voting behaviour in Bavaria makes fascinating reading. It confirms some established political trends in Bavaria, as well as revealing some new ones in both the major parties and the smaller ones, including the relatively recent phenomenon of the Republican Party.

The information was not gathered by interviews, since it was felt that interviewees can be influenced by the reactions of the interviewer. Sometimes people tell interviewers what they think the interviewers want to hear, rather than their real opinion. For this reason anonymous, written (eight-page) questionnaires were used. A great deal of detailed information

---

[1] Die Landtagswahl 1990 in Bayern. Akademie Report. Hanns-Seidel-Stiftung e.V. Munich. September 1991. The research report was compiled in the Dr Schumann Research Institute in Bruckmühl during October and November 1990.

was analysed in connection with precise informaton on all Bavarian communes and boroughs in 1990, provided by Forschungsgruppe Wahlen (the Mannheim polling institute).

Still a strong link between Roman Catholic voters and the CSU

The first key point to emerge from the survey is a clear correlation between Bavarian communes (*Gemeinden*) with a high number of Roman Catholics and support for the CSU. Such communes were shown to be a negative factor in terms of support for the SPD, as well as the FDP and the Greens.

Not only did it again emerge in this study, as expected, that Catholics in Bavaria are much more likely to give their votes to the CSU than non-Catholics, but it was also clear – and this was not necessarily expected – that living in a predominantly Catholic commune also increases the tendency to vote CSU amongst both Catholics and non-Catholics, compared with the voting habits of Bavarians living in a predominantly non-Catholic commune. At the 1990 Bavarian state elections around three quarters of those voting CSU were Catholic, although "only" around two thirds (67.2 per cent) of Bavaria's population is Catholic, according to the 1987 census.

For most of the post-war period Bavaria's population was almost exactly seventy per cent Catholic, but the figure dropped just slightly over the last decade, mainly as a result of the influx of a large number of foreigners before the 1993 change in Germany's asylum law. Its geographical position meant that Bavaria absorbed a big share of those coming from the former Yugoslavia, Czechoslovakia, Hungary, Rumania etc., as well as the former GDR.

A gradual secularisation process in west German society over many years has been another factor in the reduction in the number of Germans who are nowadays members of either one of the two main Christian churches in Germany. This factor has affected Bavaria to a certain extent, although it has been less significant there than in other parts of Germany.

Denomination has always been, and on the basis of the figures in the 1990 research still is, one of the major factors in determining electoral behaviour in Bavarian politics. An important part of this factor is the question of regular[2] church attendance, not just simply being a member of one of the two main denominations.

In fact the liklihood of a member of the Bavarian electorate voting for the CSU increases appreciably according to whether that person, either Catholic or Protestant, is a frequent church attender (of these 80 per cent vote CSU but only 10 per cent vote SPD), an occasional attender (50 per cent CSU – 18 per cent SPD) or hardly ever attends church (27 per cent CSU – 22 per cent SPD). The question of regular church-going Protestants, as well as Catholics, being more likely to vote CSU, has been a significant cleavage in electoral patterns in Bavaria since about 1970 and was confirmed by the 1990 research.

It is important to remember too that there is a marked difference in the percentage of regular church attenders in the old federal states between the two main denominations anyway. The percentage of regular church-goers in Bavaria is higher than the average for west Germany, which is approximately one third for Catholics and one tenth for Protestants. Ever since the establishment of the Federal Republic in 1949 Catholics who attend church on a regular basis have significantly outnumbered regular church-going Protestants (see Introduction).

By way of comparison, the 1990 Bavarian survey gives the figures for regular church attendance in the former West Germany as about 12 per cent, but double that (25 per cent) in Bavaria. For occasional church attendance the figures were 36 per cent overall and 41 per cent for Bavaria. Whilst over half the population of the FRG before German unity attended church either not at all or only very rarely – perhaps only for a family confirmation service or a baptism, wedding or funeral – in Bavaria this applies to a smaller proportion (around thirty-five per cent) of the population.

Whilst approximately 43 per cent of the 1990 CSU voters surveyed were regular church attenders, only 16 per cent of SPD voters were. The potential electoral support for the CSU in Bavaria is therefore very high from the outset, given that approximately two thirds of the population are

---

[2] In this context the term "regular" is taken to mean every, or almost every Sunday. See the above-mentioned report, p.13. "Kirchgang jeden oder fast jeden Sonntag."

Catholic, and that more than three times as many Catholics as Protestants are regular chuch-goers.

Although slightly more SPD voters (38 per cent) than CSU (32 per cent) voters were occasional church-goers, around 45 per cent of SPD voters never go to church compared to about 25 per cent of CSU voters. Regular church attendance is, however, the crucial factor in CSU support.

These figures confirm what has generally been the case in the "old" Federal Republic since the secularisation process began in the Fifties, namely that regular church attendance has dropped significantly but is still much higher amongst Catholics than Protestants. They also underline the special circumstances which obtain in Bavarian society, especially with regard to religion and denomination, and the special position of the Catholic Church in Bavaria.

The general link between Catholics and support for the Union parties in German elections, and the more particular connection between regular church attendance by either Catholics (more commonly) or Protestants and voting CDU/CSU has always been particularly marked in the case of the CSU in Bavaria. This is confirmed by the 1990 statistics.

During the Seventies and early Eighties, when the CSU was in opposition in Bonn, the Bavarian CSU began to make inroads into the predominantly Protestant electorate of Upper and Central Franconia, normally strongholds of the SPD, and to a more limited extent, the FDP. Again the crucial factor was regular church attendance, regardless of which denomination, and this led to voting for a party with the word Christian in its name.

SPD support in Bavaria

As has been shown already, the SPD in Bavaria is different from SPD organisations in other *Länder*. Even after making allowances for the hegemony of the CSU in Bavarian politics – which obviously has a marked effect on all the other parties in Bavaria – the SPD in Bavaria, known as the *"königlich-bayerische SPD"* is in a category of its own.

The survey revealed that in Bavaria only 35 per cent of manual workers are trade union members, compared with 47 per cent for west Germany as a whole. In Bavaria voters were asked in an exit poll in December 1990 about their feelings towards trade unions. Of those who felt a "close" connection with trade unions, 29 per cent voted SPD; however 40 per cent of those voters supported the CSU.

Although trade union membership amongst manual workers in west Germany as a whole always has been – and still is – linked to a high probability of voting SPD, as Paterson points out[3], trade union membership amongst manual workers in Bavaria is more than ten per cent lower, and the proportion of regular church attenders, linked to a high probability of voting CSU, is over ten per cent higher than the average for the former FRG[4]. The latter is a more influential factor in electoral cleavages in Bavaria.

This means that, whilst more Bavarians who feel close to trade unions support the SPD (29 per cent) than those who feel only a moderate (13 per cent) or small connection (9 per cent), the link is much weaker than in some other regions of Germany. Again Bavaria – in this case the Bavarian SPD – is an exception to the rule.

Whilst twice as many people who work in the public services in Bavaria vote SPD (around 30 per cent) as the remainder who were questioned, more than 45 per cent of them vote CSU. In the non-public service sector, however, only 13 per cent vote for the SPD, as against 53 per cent for the CSU.

The other factors which encourage SPD voting were shown to be communes with a high number of salaried employees (*Angestellte*) – more so than manual workers in Bavaria – a high population density and a high number of foreigners. Although foreigners do not have the vote, the electorate in Bavarian communes with a high amount of non-Germans tend to support either the SPD, or more probably, the FDP and especially the Green Party. The Greens do particularly well in communes with a high foreign population, which are definitely a negative factor in terms of CSU votes.

Another key electoral cleavage in Bavarian voting patterns, again confirmed by the results of the questionnaires, is the agricultural dimension. Those whose work falls in the farming and agricultural sector show a strong

---

[3] "Electoral support for the SPD by manual workers is very strongly associated with membership of a trade union." William E. Paterson and David Southern. Governing Germany. Blackwell. Oxford. 1991. p.178.

[4] According to a survey of over 20,000 Germans from the old federal states carried out by Forschungsgruppe Wahlen in December 1990, quoted on p.13 of the LTW Bayern 90 report for the Hanns-Seidel-Stiftung, already referred to.

tendency to vote CSU; the SPD, FDP and the Greens all receive below average electoral support from this section of the Bavarian population.

The survey also examined Bavarian communes where a combination of three factors was considered: population density (the number of inhabitants per square kilometre), the proportion of manual workers (*Arbeiter*) and a high or low number of Catholics. The CSU of course performs best wherever there is a high number of Catholics, but best of all where this is combined with a low population density and a high number of manual workers.

The SPD does best where a low number of Catholics is combined with a high number of manual workers – although the link between *Arbeiter* and voting SPD is a weaker one in Bavaria (just as many manual workers in fact support the CSU) – and a high population density. The FDP, whose voter recruitment potential in Bavaria is shown by the survey to be nowadays extremely weak overall, and the Greens both gain their best results in communes with few Catholics, few workers and a high population density.

Level of education

In terms of level of education of the voters, the CSU polls best amongst voters with low qualifications – a simple school leaving certificate (*Hauptschulabschluß*). Voters with slightly better educational qualifications are more likely to support the SPD; those with the highest level of education are more likely to be supporters of either the Liberals or, nowadays, the Greens.

The liberal party derives few votes in Bavarian communes with a high proportion of manual workers or people with only *Hauptschulabschluß*. The Liberals have regularly recorded worse election results at both state and federal level in Bavaria than any other *Land*. They have often struggled to clear the five per cent hurdle in Bavaria, and not always succeeded. Those who *do* support the Bavarian FDP (1990: 5.2 per cent) tend to have a high level of education[5]. This is also true of those who support the Bavarian Greens (1990: 6.4 per cent), who are favoured much more by voters in the younger age groups, and by very few over fifty.

---

[5] In 1994 at the Bavarian state election the FDP recorder its worst ever result of 2.8 per cent.

The Republicans

Falter and Schumann, in a 1988 study of contemporary right-wing radicalism, referred to four elements, which partly overlap, that are often present in areas which offer electoral support to right-wing extremist parties like the Republicans (*Republikaner*):

1. A mainly latent, but sometimes overt, expression of neo-Nazism, and the exculpation and rehabilitation of National Socialism;
2. The invocation of traditional German values – diligence, law and order, nationalism, and anti-EC feeling;
3. The harnessing of specific causes of dissatisfaction – for example, the size of the immigrant population (anti-foreigner feeling), housing and related social problems, economic difficulties, all issues typically adopted by right-wing protest parties;
4. The expression of a general disillusionment with established parties and their performance – a diffused protest symptomatic of the process of electoral dealignment.[6]

There is less data in the first part of the Bavarian report on the 1990 questionnaires, dealing with the analysis of Bavarian communes (*Gemeinden*) concerning the Republican Party (REPS), since the profile of this party and its supporters are described as possessing less clearly identifiable characteristics at commune level in socio-structural terms.

Nevertheless a number of other points did emerge clearly from the later part of the Bavarian study. Twice as many men (six per cent) as women (three per cent) vote for the REPS, they gain support from Bavarian voters living in areas with a high number of foreigners[7] (predictably) and

---

[6] Quoted from Falter, J.W. and Schumann, S., "Affinity Towards Right-wing Extremism in Western Europe,"in West European Politics, 11 (2), April pp.91-110, by Gordon Smith in his chapter Dimensions of Change in the German Party System, pp. 92/93. This is a contribution in Stephen Padgett. Parties and Party Systems in the New Germany. Dartmouth. Aldershot. 1993.

[7] "The success of the Republicans is due in large part to their ability to exploit the issue of foreign workers, asylum seekers, and German workers from the East whose repatriation, they argue, has been encouraged by lavish state subsidies and social benefits." Stephen Padgett. The Party System, in Developments in West German Politics, ed. by Gordon Smith, William E. Paterson, Peter H. Merkl. Macmillan. London. 1989, p.145.

their voters, like those of the CSU, usually have lower educational qualifications (they are often manual workers).

People working in the general area of trade, insurance and the media in Bavaria are more likely to give their vote to the *Republikaner* than are civil servants and salaried employees. It must be pointed out, however, that all these characteristics appear as minute indicators in the research statistics, since the profile of the Republican Party is much smaller overall.

Of those voters who considered the question of "foreigners" to be the most important election issue, 20 per cent voted for the Republicans, whose overall vote in the 1990 Bavarian state election was only 4.9 per cent. Although the REPS were the second most popular party in this context (foreigners), the CSU gained 50 per cent here (overall vote 54.9).

When questions were put on the image of the various parties, the "friends"[8] of the Republicans associated the party with caring about both "ordinary people" and the question of internal security, although it should be added that, unlike the Greens and the FDP, respondents thought the Republican Party showed commitment in all areas, except that of representing convinced Christians.

Although previous surveys of support for the REPS (e.g. Spiegel, Forschunsgruppe Wahlen) in other areas of Germany had suggested more Protestants than Catholics – disregarding the question of regular church attendance – vote for them, the special circumstances in Bavaria mean that around 75 per cent of the few who do support the REPS – incidentally the same is true for the Bavarian Greens and Liberals too – are nominally Catholic. Only about 15 per cent of Republican supporters are regular church-goers, however. Regular church attendance is again the crucial point when discussing the denominational dimension in Bavaria.

When "friends", i.e. anyone whose reaction to a particular party was generally positive, of the small parties were asked to place themselves on a left-right scale, 49 per cent of the friends of the REPS, 36 per cent of the friends of the FDP, and 10 per cent of the friends of the Greens considered themselves to be on the right. Whilst the majority of FDP and Green

---

[8] In the survey a distiction was made between "friends" of a party – those whose answers were positive regarding that party – and "opponents," i.e. those whose responses were negative. See page 25 of the report.

friends placed themselves in the centre of the political spectrum, nearly as many Green supporters (almost 40 per cent) considered themselves to be on the left.

Nearly 40 per cent of those who were antagonistic towards foreigners thought the Republicans were the party most competent to deal with the problem. Almost as many people thought the CSU was as competent, but the other parties were well behind on this issue. With regard to the voters' image of the parties in Bavaria, the REPS challenge the CSU on only one quality – toughness.

Around 28 per cent of the friends of the REPS are manual workers and approximately 18 per cent have a positive attitude towards trade unions. In terms of age, the REPS do best amongst Bavarians in the 30-50 category (45 per cent). Their second best age category – although a long way behind (20 per cent) – is the over-60 group. In Bavaria the under-30 age group, normally considered a potential source of major recruitment for the Republican Party in Germany – particularly amongst the young unemployed, gave, perhaps surprisingly, only 18 per cent support to the REPS in 1990, according to this report.

More predictable results, however, were forthcoming in other respects. Areas where the Republicans scored high amongst their supporters in this study were in four questions connected with native surroundings (*Heimat*) and foreigners (*Ausländer*). Friends of the REPS felt an extremely close connection with Bavaria[9] (*Bayern verbunden*) – 98 per cent – and the same percentage thought that foreigners were *not* an enrichment of their culture. Sixty per cent of Republican supporters were proud to be German and 92 per cent were in favour of sending foreigners home. In all four questions the figures for the REPS were considerably higher than those for the FDP, whose results were in turn higher than the Greens in all four questions.

---

[9] This is particularly interesting, as the Republican Party is essentially a Bavarian party, founded in 1983 by three former CSU politicians (Handlos. Voigt and Schönhuber), although not all Bavarians are keen to recognise this. In an interview with the author at the Bavarian Staatskanzlei in Munich on 14.10.1993 Dr Baer emphasised that the Republicans were in the state parliament of Baden-Württemberg but not Bavaria (4.9 per cent in 1990). In the event they also failed to clear the five per cent hurdle again at both the Bavarian and federal elections in 1994.

REPS supporters came out very strongly in favour of wanting clearly formulated conditions, as opposed to continual change, peace and order rather than movement and innovation, tradition instead of change, and fixed rules rather than improvisation. In this sense, the view has sometimes been expressed that if a challenge to the CSU as the *"Staats- und Ordnungspartei"* in Bavaria ever came, it would be most likely to come from the Republicans[10]. The CDU/CSU is undoubtedly vulnerable to a loss of support if parties like the Republicans can increase their electoral appeal[11]. Although the actual result of the October 1994 federal election in fact brought the Republicans little success (less than two per cent), it was not until shortly before the national vote in September 1994, when the party fell below five per cent on its home territory of Bavaria (3.9 per cent), that any political pundits dared to forecast a result of less than five per cent in Bonn.[12]

At the beginning of super election year (*Superwahljahr*) 1994, with seventeen local and regional elections in Germany, plus the European, federal and presidential elections, the Republican leader, Franz Schönhuber, announced that his party would be contesting every election. The fortunes of both the the party and its leaders fluctuated during the course of the year.

Even though the EMNID polling institute's forecasts at the beginning of 1994 (mid-January) continued to predict only between 2 per cent (West) and 3 per cent (East) support for the REPS in a federal election[13], the party had caused enough surprises in state elections already. Its result of 10.9 per cent in Baden-Württemberg in April 1992 was just one example of this.

Despite the fact that the Republican Party was not represented in the Bavarian parliament following the 1990 state elections (4.9 per cent), there has been evidence from time to time that they seek to occupy a political position between the "old" extreme right of the German People's Party (DVU) and the National Democratic Party (NPD) on the one hand, and the

---

[10] See footnote 6.

[11] op. cit. Gordon Smith, p.93.

[12] In interviews already referred to – with Alf Mintzel in Passau and Wolfgang Gibowski in Bonn – the author was told that this was a controversial issue with arguments on both sides.

[13] EMNID poll predictions reported on the German TV news channel N-tv, 16.1.94.

CDU/CSU, on the other, as Saalfeld points out[14]. The REPS, consequently, are definitely targeting some of the CSU's voting potential in Bavaria, whether the CSU is prepared to admit it or not. This has been the case to some extent with the federal CSU ever since the CDU/CSU returned to power with the FDP in 1982/83 and opted for the strategy of trying to conquer the middle ground of German politics. That left an opening on the right, which the REPS have not been slow to fill.

The party of non-voters –
a phenomenon of the Nineties

Many observers of German politics have been concerned with the trend in recent years towards lower turn-out figures in elections and the feeling amongst the German electorate of *Politikverdrossenheit* – a growing disillusionment with politics generally and a lack of confidence in the country's main parties and their leaders.

One part of this phenomenon can be traced back to several financial scandals in the 1980s, of which the Flick Affair was the most notorious. That particular scandal tainted all the main political parties, except the Greens. Germans who consequently lost confidence in the political system first coined the phrase *Staatsverdrossenheit* – the state "gone stale."

A new stage in the development of this phenomenon came about after the trough of tremendous disappointment amongst many ordinary German citizens, which followed the hysterical celebrations of the fall of the Berlin Wall in November 1989 and the euphoria that preceded and accompanied the celebrations of German Unity and the first all-German elections in late 1990.

In 1991 and 1992 the true costs of reunification, in both financial and social terms, began to make themselves felt in the cold light of day, after the party was over. The widespread feeling of helplessness and having been badly let down, which followed Helmut Kohl's pre-election promises that no German would be worse off and that there would be no tax increases

---

[14] See "The Politics of National-Populism: Ideology and Policies of the German Republikaner Party" by Thomas Saalfeld in German Politics vol. 2 No. 2 1993, pp.177-199.

after reunification, contributed to a general disillusionment with leading politicians and politics in general. That applied particularly in east Germany, where people's expectations were very high – perhaps unrealistically high – but in the west too, especially since many west Germans are now financially worse off than before. Many resent having to pay for their "brothers and sisters" in the east.

In Germany during the approach to super election year there was a pronounced feeling that the country's leading politicians had lost touch with the things that mattered most to ordinary citizens. This disillusionment, or *Politikverdrossenheit*, has manifested itself frequently at elections in recent years in the form of abstentions and protest votes.

Many of the EMNID opinion poll surveys during 1993 on the so-called *Sonntagsfrage* (which party would you vote for if there were a federal election next Sunday?) reached the conclusion that the largest party was *die Partei der Nichtwähler* (the non-voters' party). This general mood of disappointment and disillusionment with the German political system prevailed throughout 1993 and was still in evidence in January 1994 when the EMNID poll again asked which party could best solve Germany's current problems. The result, not for the first time, showed that 26 per cent thought the SPD, 22 per cent the CDU/CSU, but 39 per cent said they thought that no party could solve the problems.

In the 1990 Bavarian survey the number of abstainers increased in all age groups amongst both men and women vis-a-vis the 1986 Bavarian state election. Increases were in the order of 2-3 per cent for both sexes in the older age groups and 6-8 per cent amongst the 18-35 group.

A particularly interesting aspect of the research in Bavaria also confirmed a widespread feeling of *Politikverdrossenheit*. When voters were asked whether they felt that politicians took an interest in what ordinary people think, only just over 16 per cent replied "yes", with over 83 per cent saying "no". Remember that this information was gathered in October and November 1990 before the general mood of optimism had started to change.

In the same investigation over 64 per cent of those questioned answered in the affirmative to the question: "Is politics is a dirty business?", with only 38 per cent saying "no". Roughly 60 per cent of those questioned felt that politics was a dirty business, as well as feeling that politicians were out of touch with the concerns of ordinary people. This seemed to be a very high level of anomie with the political process (*Politikverdrossenheit*), which of course provides an ideal breeding ground for extremist parties of either the Right or the Left.

There was no significant difference between men and women or Catholics and Protestants on this question, but diferences did occur in two categories. First, a gradual increase in *Politikverdrossenheit* is evident as age decreases. The voters in Bavaria who feel most disillusioned with politics were in the 18-34 age group, the next most disillusioned 35-44, then 45-59. The least disillusioned (but still with a figure of over 52 per cent) were the over-sixties.

Second, by far the most disillusioned group were those who attend church rarely or never (74 per cent), whereas those who attended church regularly were less dissatisfied with the political process. As far as party support in Bavaria is concerned, those who vote for the CSU experience far less anomie/disillusionment than supporters of other parties.

When those who had not bothered to vote in the 1990 Bavarian LTW were asked why they had not voted, the reason most frequently given (28 per cent) was "false election promises," followed by a feeling of having "no political influence" (12 per cent) and the view that no party was either "completely satisfactory" in their view or "electable" (11 per cent each).

The same people were asked whom they would elect at a federal election, and those who replied "no party" were also asked why. "False election promises" again came top of the list (29 per cent), with 16 per cent citing either "politics is a dirty business" or "no party electable." The fourth most frequent reason given for not voting was as a protest against "reunification" (14 per cent).

On the other side of the coin, the three most common reasons given *for* voting for a party – and here the CSU was well ahead – were: "good achievements so far" (69 per cent of CSU voters mentioned this but only 15 per cent of voters of other parties), "the party has good politicians" (CSU 57 per cent – others 16 per cent), and "the party is trustworthy" (CSU 37 per cent – others 19 per cent).

CDU candidates in Bavaria?

Another interesting aspect of the 1990 Bavarian LTW study was the matter of whether the CSU should consider putting up candidates outside Bavaria and whether Bavarian voters would vote for CDU candidates if they stood for election in Bavaria. Almost seventy per cent of those questioned said they would not have voted CDU at the 1990 Bavarian state election, had this been possible. Only fourteen per cent answered "yes" and eighteen per cent did not know. This was the first time this question had been put in relation to a regional election, although it had been put in relation to a federal election.

Of those who said they would have voted for a CDU candidate, the vast majority (over 70 per cent) were CSU voters. Thirteen per cent of REPS voters, seven per cent of FDP voters, two per cent of SPD voters and seven per cent of those who did not vote for any party also said they would have supported a CDU candidate.

The only time when there was ever any serious talk of the CSU campaigning outside Bavaria was in the autumn of 1976 in the aftermath of the federal election. The CSU met in Wildbad Kreuth in Bavaria for a special party conference and, under the chairmanship of Franz Josef Strauß, there were proposals for the CSU to break off the long-standing parliamentary alliance with its sister party, the CDU.

Polls taken at the time showed that the formation of a CDU association (*Landsverband*) in Bavaria would have split the conservative vote, endangering the CSU's omnipotent position (*Vollmacht*) on its home territory. By the time the new Bundestag was convened, towards the end of 1976, the CDU/CSU differences had been patched up and the two parties again formed a joint parliamentary party (*Fraktionsgemeinschaft*), which many would say has been the secret of the success of the two German conservative parties since 1949.[15]

The development in 1946 of an independent, separate CSU association for the Land of Bavaria, rather than allowing the CDU to establish a party

---

[15] See an article by Clay Clemens, "The CSU and West German foreign policy in Politics and Society in Germany, Austria and Switzerland, Vol. 2, No. 1/2, Spring 1990, p.20.

association there, as it did in each of the other *Länder*, is viewed by many as one of the fundamentally significant achievements of the party's founding fathers. There has been no evidence at any time since 1946 that it would be in the CSU's interests to campaign outside Bavaria. Nor would it be in the interests of the CDU to put up candidates inside Bavaria. The findings of the 1990 study simply confirmed this state of affairs.

Most important problems for the voters

There is evidence that nowadays the main campaign issues in state elections are often related closely to the national issues that are important in federal elections[16]. Also, the proximity of the state election in Bavaria (14 October) and the federal election (2 December) in 1990 clearly influenced the voters' choice when they were asked in the questionnaire about key issues.

Voters were asked to state in order the three most important problems in the FRG which needed to be solved. It was felt that this method would produce a more accurate picture than presenting the respondents with a list of topics from which they had to choose.

The problem mentioned most frequently (by nearly 30 per cent) was the environment. Just under 20 per cent identified domestic policy for (the new) Germany (*Deutschlandpolitik*) and the third most frequently mentioned topic was aliens (*Fremde*) (16 per cent). Whilst the Green Party would surely claim some credit for maintaining the environment as an important issue in the forefront of people's minds, some would argue that the Republicans were partly responsible for the Bavarian voters' concern with aliens.

After these three came unemployment (13 per cent), housing problems (8 per cent) and the economy (6 per cent). All the other problems named, except pensioners/old people and social policy (2 per cent each), were men-

---

[16] This was confirmed in an interview during the Rundschau news on Bavarian television (Bayerischer Rundfunk) on 20.01.94 by Prof. Heinrich Oberreuter, the director of the Politische Akademie in Tutzing, Bavaria. In connection with considering likely Bavarian election campaign themes in 1994, Oberreuter said that the balance in the German federal system had been tilted more in favour of the federal than the regional level, and that key national issues such as unemployment, internal security, reunification and the like were frequently the main issues in state elections too.

tioned by one per cent or less, including a policy for women (*Frauenpolitik*), at the very bottom of the list.

The report then looked only at the topics which had been named as one of the three most important. Over half of these respondents thought that the environment was one of the three most pressing problems. The explanation for this was not sought in the study, but it may well lie in the fact that Bavaria has suffered from pollution on its eastern borders on several occasions, the worst of which was the Chernobyl disaster in 1986, when Bavaria was the worst affected of the German states, owing to its geographical position.

Bavaria has always been very sensitive towards the protection and preservation of its many forests, lakes and beautiful scenery. Many Bavarians feel an affinity with nature as an integral part of their upbringing and culture, which goes far beyond simple environmental protection measures.

In comparing voting patterns with problems named, there are significant differences according to the specific problem. This feeling was most marked in the case of those who named unemployment amongst the three most pressing problems. Here 62 per cent voted CSU (the highest figure). Over half of those who listed *Deutschlandpolitik*, the economy and the environment voted CSU, but only 35 per cent of those who mentioned social policy did so; 28 per cent of them voted for the SPD.

When problems such as pensioners/old people or housing cropped up, even more people voted SPD, to the detriment of the CSU. It is interesting to note that the Republicans also gained more votes from people who mentioned these topics – and also of course the issue of foreigners – which confirms the impression given in other findings in the report, namely that the REPS are seen by their own supporters as a type of protest party, as far as social problems are concerned.

More people who included the environment amongst the three most important problems gave their votes to the Greens, of course, than with other issues. Amongst all the other problem areas mentioned, in fact, there were more Green voters who did not name the topic than did. Further analysis showed that amongst those who referred to the environment as *the* most important issue, 17 per cent voted for the Green Party (the second highest figure behind the CSU with 46 per cent), but on average only 7 per cent did so.

Similarly, in the case of those who mentioned foreigners as the most important problem, 20 per cent voted for the Republicans (the second highest figure behind the CSU with 50 per cent), but on average only five per cent did so. Respondents who thought that foreigners did not represent an enrichment of their culture, as well as thinking that as many foreigners as possible should be sent home, were asked to state which party was closest to representing their views on the problem of foreigners.

An above average number of those who in the survey had voted for either the CSU or the REPS had said "yes" to the two questions above, but the reverse was true for those who had cast their votes for the SPD, the FDP and the Greens.

Although the time when the data for this research were collected (October and November 1990) was too soon after German unity on 3.10.1990 for too much significance to be attached to the findings, they certainly are of interest and shed some light on electoral behaviour in Bavaria. It should also be remembered that frenzied debates on the reunification of Germany had been going on since shortly before the fall of the Berlin Wall in November 1989.

Of those who were in favour of German reunification, 31 per cent believed that the CSU was the party which best represented their interests, but of those who were against it, only 11 per cent thought so. This is hardly surprising, as the CSU was also a member of the federal government at the time of reunification. Forty per cent of those in favour thought that the CDU best represented their interests, and 45 per cent of those against reunification (or felt it had occurred too soon) considered that the SPD best represented them. Just over 20 per cent of those in the survey who were against reunification thought that *no* party represented their interests.

Overall, about 60 per cent of all those questioned in this poll were in favour of reunification and the vast majority of those who were in favour (70 per cent) thought their interests were best served by either the CDU or the CSU. Around 35 per cent either wanted German Unity at a later date or rejected it entirely for financial reasons. However it must not be forgotten that these questionnaires were completed in the immediate aftermath of reunification, before the 1990 first all-German federal elections, at a time when a generally euphoric atmosphere, for the most part, still prevailed.

The Left-Right dimension on the Bavarian political spectrum

The questionnaires in this poll required the respondents to place themselves, politically speaking, in one of eleven squares on a left/right spectrum. Of those who judged themselves to be on the right of the political spectrum, 76 per cent vote CSU and 8 per cent REPS. These were the only two parties with more voters on the right than the centre or left. In fact the REPS were the only party with no supporters at all who categorised themselves as being on the left of the spectrum. The CSU had 45 per cent support from the centre and ten per cent from the left.

The vast majority of those who see themselves as being on the left vote either for the SPD or the Green Party, in approximately equal numbers (about 30 per cent each). Only 2 per cent of those on the right and 4 per cent of those in the centre vote for the Greens, but 20 per cent of those in the centre politically support the SPD. As far as the FDP is concerned, the self-assessment spectrum makes very little difference, as they struggle to clear the five per cent hurdle, although most friends of the Liberals saw themselves in the centre ground, and more on the right than the left.

Summary

In summarising the main electoral cleavages in Bavaria, the study examined the ties which exist between the supporters of the two major parties, the CSU and the SPD, and the way they determine how people vote, under five headings. Firstly, denomination: 75 per cent of CSU voters and 53 per cent of SPD voters in Bavaria are Catholic.

Secondly, church attendance: 43 per cent of CSU voters but only 16 per cent of SPD voters are regular church-goers. Occasional attendance at church applies to 35 per cent of CSU voters and 39 per cent of those who vote SPD. Amongst CSU supporters 25 per cent attend church rarely or never; this is true for 45 per cent of SPD supporters.

Thirdly, manual workers: the percentage of manual workers (*Arbeiter*) amongst those who vote for the CSU and the SPD is almost identical (just over 40 per cent). This factor is therefore, unlike in most areas of Germany, **not** relevant in determining electoral behaviour in Bavaria.

Fourthly, trade unions: amongst CSU voters 23 per cent, amongst SPD voters 53 per cent view trade unions positively. Trade union membership is less relevant in influencing how people vote in Bavaria than elsewhere in Germany, where it is normally still a reliable indicator of SPD support.

Fifthly, self-assessment on the political spectrum: a mere two per cent of CSU voters place themselves on the left, compared with 65 per cent of SPD voters. Around 45 per cent of CSU and 65 per cent of SPD voters see themselves as being in the centre. Most CSU supporters assess themselves on the right (52 per cent), but only 10 per cent of SPD voters do this.

Finally, in a comparison of the SPD and CSU, the study looked at three areas briefly: feelings towards home and foreigners, cognitive orientation of voters and social integration versus disillusionment.

In the first of these categories significantly more CSU voters felt a strong connection with Bavaria – a staggering 98 per cent in fact – than did SPD voters (67 per cent) and just over 60 per cent of CSU voters reported that they were proud to be German, but only just under 40 per cent of SPD voters did so. More CSU voters (73 per cent) than SPD voters (61 per cent) were in favour of sending as many foreigners as possible home, and over 80 per cent of CSU voters and over 60 per cent of SPD voters thought that foreigners were not an enrichment of their culture.

In what it called the cognitive orientation system – various preferences were expressed by voters on whether their natural sympathies were closer to clear-cut arrangements, peace and order, fixed rules, tradition and the like. In all nine questions – in general terms whether the voters questioned preferred a more traditional approach or plenty of reform and change – CSU voters voted to a greater extent than SPD voters in favour of preservation of the status quo. For example, 82 per cent of CSU voters and 62 per cent of SPD voters preferred fixed rules to improvisation.

Finally, in terms of social integration, CSU voters thought, in significantly greater numbers than SPD voters, that it was still possible to enjoy a peaceful, secure life-style, that most people still had a firm base and that the world was still reasonably comprehensible, where it was possible to keep track of events. In terms of disillusionment with politics, more SPD voters (98 per cent) than CSU voters (81 per cent) felt that politicians were not interested in the ordinary citizen; 44 per cent of CSU voters and 79 per cent of SPD voters said that politics was "a dirty business."

9

Franz Josef Strauß – an Extraordinary Phenomenon

Franz Josef Strauß (1915-88) was without doubt the most popular and successful post-war Bavarian leader; at the same time, he was arguably also the most controversial politician in the Federal Republic. Some considered Strauß a pragmatic politician who believed in taking the art of the possible right up to (and in some cases almost beyond) its limits. No German politician, with the possible exception of the late Willy Brandt, aroused the passions and emotions of the West German electorate to the same extent as Strauß.

However, the German voters were prepared to accept Brandt as Federal Chancellor in 1969 and 1972 – he even survived, by a whisker, the first attempt in the FRG at a constructive vote of no confidence in the Bundestag – but they were *not* prepared to support FJS in his efforts to become Chancellor in 1980.

Strauß was active in both Bavarian and German politics from the very beginning of the postwar period (*ein Mann der ersten Stunde*) right up until the day he died in the autumn of 1988. Since he was such a colossal influence on politics in the Federal Republic for over four decades, with his idiosyncratic approach, especially in the last decade of his life, when he was simultaneously minister president of Bavaria and chairman of the CSU, an analysis of the man and his role is required, in order to understand fully the special position of Bavaria in the post-war period, as well as in the new Germany. The enormous influence of Strauß on Bavarian politics in general and the CSU in particular is still clearly evident today.

Early influences

FJS was born in Munich on 6 September 1915, the second child of Walburga, nee Schießl, and Franz Joseph Strauß, a butcher. The young Franz Joseph – he was named after his father and later adopted the more modern spelling Josef – was brought up with his elder sister, Maria, in an environment which Strauß himself described in his memoirs as "Catholic, monarchist and anti-Prussian"[1]. In 1922 he started primary school in Munich's Amalienstrafle, followed by one year at secondary school (*Realschule*) before transferring to grammar school (*Gymnasium*). At the age of eighteen, Strauß became Bavarian cycling champion, a title of which he was very proud, even in later years.

After passing his "A levels" (*Abitur*) in 1935 – his results were amongst the best in Bavaria in that year – and completing six months' community work (*Arbeitsdienst*), Strauß began to study German language and litera-ture, classical languages and history, (plus some economics later) at Munich university in 1936. One year later he joined the National Socialist drivers' corps (NSKK). In 1939 Strauß was called up for the army (heavy artillery). First he went to Landsberg and was then transferred to Trier. In 1940 FJS passed the first part of his examinations to be a school teacher (his-tory and classics) and then took part in the French campaign as a corporal. After passing the second part of his university qualifying examinations during the following year, soldier Strauß participated in the German army's harsh Russian campaign.

As part of the air defence unit Strauß spent time in Denmark, France and the Ruhr. In August 1942 he was stationed near Schongau in Upper Bavaria, and one year later he received his official approval as a grammar school teacher. On 2 June 1945 the US military government made Strauß "assistant" *Landrat* (chief administrative officer) in Schongau; in November of that year he founded, along with Andreas Lang and Franz Xaver Bauer,

[1] Franz Josef Strauß. Die Erinnerungen. Siedler. Munich. 1989, p.15.

a CSU association in Schongau. The CSU had been officially founded at local level the previous month.

The start of a career in politics

In 1946 Strauß developed contacts with Josef Müller (*der Ochsensepp*) and Fritz Schäffer, who represented the two factions of the CSU. From 6 June 1946 until 13 January 1948 Strauß worked in the youth section of the Bavarian education ministry. On 31 August he was elected *Landrat* of Schongau, and at the end of the year Strauß joined the CSU executive as a youth representative. In early 1948 he became a senior civil servant in the Bavarian ministry for home affairs. In the same year the bizonal Frankfurt Economic Council, was established. Strauß became a member and committed himself to strong support of Ludwig Erhard's social market economy. Interestingly, Erhard, although he was from Bavaria, was a Franconian Protestant from Fürth. At that time he was not a member of any political party. He did not want to join the CSU – there were actually approaches from the Bavarian FDP, owing to his stance on economic liberalism – and when he finally joined a party, only after he had become Economics Minister in Adenauer's first cabinet in Bonn, he chose the CDU.

Just before Christmas 1948 Strauß took over the office of Secretary General of the CSU, which he retained until 1953. On 14 August 1949, at the first federal elections, Strauß won the constituency of Weilheim, which contained Schongau, south-west of Munich, with 28.8 per cent of the vote. FJS retained this direct seat at every federal election, until he resigned from the Bundestag in Bonn in 1978, in order to take over as Bavarian minister president in Munich.

In the first federal parliament FJS became the deputy chairman of the twenty-four-strong CSU group. He was also a member of the executive of the CSU parliamentary party (*Fraktion*). The American High Commissioner, John McCloy, invited Strauß to join a select circle to discuss a German defence contribution. FJS initiated the foundation of a Bundestag parliamentary committee for European security and in this way became the defence spokesman for the CDU/CSU *Fraktion* in Bonn.

Strauß joins Adenauer's cabinet

Konrad Adenauer once said of Strauß that he was a clever and imaginative man who was gifted politically. In October 1953 Chancellor Adenauer gave Strauß his first ministerial appointment, but it was not straightforward. Adenauer first offered Strauß the job of minister for family affairs. FJS refused. Adenauer was apparently furious[2]. Strauß insisted that it was not the right post for him; in any case, he argued, as a bachelor he was not going to accept the job of *Familienminister*. It appeared that Adenauer preferrred to have Strauß inside the cabinet, where he could keep an eye on him, than outside it. So, at the age of thirty-eight, FJS became the youngest member of the cabinet, as one of four ministers for special tasks (*Minister für besondere Aufgaben*) – the others were Waldemar Krafft, Hermann Schäfer and Robert Tillmanns.

Their task was to liaise between the parliamentary parties in the coalition and the West German government. Strauß was unhappy with such a vague brief and decided to use his position to develop contacts abroad, particularly with France and Luxembourg. When Adenauer addressed the Saar question in 1954, he exploited the friendship between Strauß and the French foreign minister, Pinay, in order to reach an agreement.

Two years later Adenauer appointed Strauß as federal minister for atomic energy. FJS was fascinated by modern technology and revelled in the task of trying to bring West Germany's scientific knowledge and research in this new area up to the level of other countries. A nuclear reactor was built in Munich-Garching. It was in the news again in 1994, when the Bavarian Education Minister, Hans Zehetmair, was discussing plans for the reactor's modernisation and expansion, in connection with new job opportunities.

Only four weeks after taking office FJS flew to the USA to negotiate a draft agreement on cooperation over the development and exploitation of nuclear energy, of which which he remained a staunch supporter all his life. Strauß also initiated research projects on new energy sources (e.g. water and solar power).

In the early Fifties Strauß made several parliamentary speeches in Bonn

[2] Franz Josef Strauß. Der Mensch und der Staatsmann. Herausgaber Walter Schöl. Schulz. Percha. 1988, p.85.

on European security, Germany's defence policy and the question of German re-armament. He was considered to be a brilliant orator by many, including his "arch-enemy Rudolf Augstein, who desribed FJS in 1979 as the best public speaker in Germany, along with Herbert Wehner and Rudi Dutschke. Augstein repeated this view in the days after the death of Strauß[3].

Germany re-arms

In July 1955 Theodor Blank was appointed as the Federal Republic's first Defence Minister, with the task of establishing a West German army. Blank's proposals, however, proved unfeasable, and in October 1956 he was replaced by Strauß.

There were rumours that Strauß had deliberately undermined Blank's position. This was never proved. FJS himself refers to the allegation in his memoirs, stating that the assertion by Hans-Peter Schwarz (*die Ära Adenauer*) that Strauß needed six years to "overthrow" Blank was untrue, although he admits he wanted to be the Federal Republic's first Defence Minister[4]. At a time when the Cold War was just beginning, the participation of the new *Bundeswehr* in the nuclear plans of the western Alliance caused considerable controversy.

In July 1957 Strauß (aged forty-one) married Marianne Zwicknagl (aged twenty-seven), the daughter of a brewery-owner, in Rott am Inn in Lower Bavaria. She had a degree in economics (*Diplomvolkswirtin*), as well as qualifications in French and English. The marriage ceremony was performed by the arch-bishop of Munich-Freising, Cardinal Wendel, and Konrad Adenauer was a guest of honour. On the evening before the wedding fifteen army recruits drowned trying to cross the river Iller near Kempten. The Defence Minister vsited the scene of the tragedy and very nearly missed his own wedding.

Marianne and Franz Josef Strauß had three children. The two sons were Max Josef, the eldest child, now a lawyer, and Franz Georg, who went into advertising and set up the "Weiß-Blau" television station. Their daughter Monika (married name Hohlmeier) was the only one to pursue a career in politics. She was elected as one of the four deputies of party chair-

---

[3] Spiegel 10 October 1988, p.19.

[4] op. cit. Die Erinnerungen, p.268.

man Theo Waigel at the 1993 CSU party conference and was a member of the Bavarian cabinet (junior minister for education). Monika Hohlmeier is a member of the present Bavarian cabinet.

Marianne Strauß died in a car accident in 1984. Her husband was abroad at the time. It was rumoured that their marriage was going through a difficult patch and that Marianne Strauß was under some strain at the time. She was driving home alone around eleven-thirty one evening, after visiting a friend, along a good road which she knew well. Her car left the road on a bend – for no apparent reason – and hit a tree. No other vehicles or people were involved. Frau Strauß was not found until the next morning. The circumstances of the accident were never fully explained.

In March 1961 FJS was elected party chairman of the CSU, after Hanns Seidel had to step down owing to illness. Strauß retained this office for twenty-seven years, until the day he died. He was re-elected fourteen times with well over 90 per cent of the party delegates voting for him on most occasions. The highest results were 99 per cent in 1979 and 98.8 per cent in 1985. The only "dips" in the party chairman's tremendous popularity were 86.8 per cent in 1963, six months after the Spiegel Affair resignation, and 77 per cent in 1983, following his involvement in the massive loan to the GDR (*Milliardenkredit*).

The period 1956 to 1962, ending in FJS being forced to step down as Defence Minister (see Spiegel Affair below), was an extremely controversial one – especially in the eyes of Strauß's critics. This was the period when most of the political affairs occurred, with which it was alleged that the name of F.J. Strauß was connected. In the immediate post-war period FJS had said that the hands should drop off any German who ever held a weapon again. In February 1952, in a Bundestag debate, he was nonetheless one of the most enthusiastic proponents of establishing a West German army. Strauß later insisted he meant that no German should take up weapons in an offensive, not a defensive role.

Before the 1961 federal election, the FDP, under its chairman Erich Mende, took a decision to continue the government coalition with the CDU/CSU after the election if possible – but without Adenauer, who was then aged eighty-five. In the event Adenauer stubbornly refused to step down – the best the FDP could do was to get an agreement from "the old man" (nickname: *der Alte*) to retire in 1963. As a result the FDP gained the reputation of a turncoat party (*Umfallpartei*). Although the name of Strauß is not usually linked with this incident, the author was told in two independent interviews in Bavaria that at the time FJS first promised that the CSU would join the FDP in not voting for Adenauer as Chancellor after the elections, but then went back on his word.[5]

At the beginning of 1963 Strauß was chosen as chairman of the CSU *Landesgruppe* in the *Bundestag*. He dedicated his efforts towards foreign and security policy, as well as the German economy, based on the idea of a federal Europe and Franco-German links. This policy brought the CSU leader into conflict with the Foreign Minister of the day, Gerhard Schröder, who was known as an "Atlanticist". This was not to be the last occasion on which FJS trespassed in the field of foreign affairs.

Strauß as finance minister and party spokesman

Following the resignation of Ludwig Erhard and the formation of the Grand Coalition in December 1966, Strauß became federal Finance Minister. He actually worked very closely – and in fact successfully, despite the mutual distrust between FJS and the Social Democrats – with Professor Karl Schiller (SPD), the Economics Minister. They were known in common parlance at the time as *"Plisch und Plum"* – a sort of Tweedle-Dum and Tweedle-Dee. The German expression, taken from a well known children's story by Wilhelm Busch, also incorporates the idea of "partners in crime".

When Strauß died, Schiller said of him that he was a pugnacious colleague, on whom one could however fully rely, once a common line of argument had been agreed[6]. Even FJS's critics acknowledged his contribution to improving the health of the nation's finances during this period. Although it represented a return to power in Bonn for Strauß, it was generally known

---

[5] This version of events was related to the author in an interview with Ulrich Witschel, an FDP member of the Bamberg Stadtrat on 29 April 1982, with the comment: "Umgefallen ist damals Herr Strauß". In an interview with Hermann von Schaubert, at the time Head of the Thomas-Dehler-Institut, in Munich on 22. July 1983 the same version was given to the author quite independently.

[6] "Strauß war ein streitbarer Kollege, auf den man sich, wenn eine gemainsame Linie gefunden war, voll verlassen konnte." Quoted in Franz Josef Srauß. Stern Extra-Ausgabe. 7 Oct. 1988, p.96.

that he was not an enthusistic supporter of the Grand Coalition and never felt completely happy in the CDU/CSU/SPD government (1966-69).

From 1969 onwards, with the CDU/CSU in opposition, Strauß, still a Bundestag deputy, was his party's parliamentary spokesman on finance and economics. He was outspoken, blunt and frequently explosive, as he fought many passionate debating battles with the SPD/FDP in Bonn over tax and investment policy in a climate of rising unemployment during the Seventies.

In Bavaria, meanwhile, the hegemonic position of the CSU had reached oligopolistic proportions, as the party gained over 62 per cent at the Bavarian state elections in 1974 and 60 per cent of the Bavarian vote at the 1976 federal elections. In November 1978 Strauß took over from Alfons Goppel as Bavarian minister president, resigning his seat in the *Bundestag*. He retained this post for ten years, until he died.

The controversial Chancellor candidate

Strauß now set his sights on the position he had always wanted. After defeating Ernst Albrecht of the CDU, he was nominated as the Union parties' chancellor candidate for the 1980 federal election. This polarised the West German electorate to a unique extent. Groups in German society never before heard of began to support either one of the two candidates for Federal Chancellor. Strauß, who had been extremely critical of Helmut Kohl's unsuccessful attempts to unseat the SPD/FDP federal government in 1976, was determined that for the first time since 1949 the joint CDU/CSU candidate for Federal Chancellor would be a Bavarian and a member of the CSU.

The 1980 election campaign was emotionally charged. Artists and intellectuals canvassed for Helmut Schmidt (SPD), whilst a variety of workers' and women's groups and top sportsmen pledged their allegiance to Strauß. In Helmut Schmidt, however, FJS faced a formidable opponent – probably the most gifted and competent Chancellor since Adenauer, and in the opinion of many simply the best in the history of the FRG.

The nub of the problem for Strauß was that he aroused equally fierce resentment outside his home state – and particularly outside southern

Germany – as he did passionate support inside Bavaria. At an election rally in the Ruhr, for example, he had to speak from behind a bullet-proof screen and was pelted with eggs. In an uncontrolled outburst, Strauß told his predominantly young, hostile audience that they would have been the best disciples of Goebbels. The remark was not well received in the German media.

Unfortunately for Strauß and the CSU, many CDU members and voters north of the river Main found FJS's political style and rough-and-ready manner simply unacceptable; they preferred to put their trust in the incumbent SPD Chancellor, Helmut Schmidt, whom they felt was a better choice to represent their country at home and promote its image abroad. For many CDU supporters in northern Germany, Schmidt was the best Chancellor the Federal Republic had ever had – but simply a member of the wrong party.

The schism in the West German electorate had polarised people into one of two groups – you were either for Strauß or **against** him – no half measures. Few people felt indifferent towards FJS. Posters, stickers and badges were sold bearing the words *Stopp Strauß* and *Strauß – nein danke*. The divided mood was reflected in the election result – almost an even split between the parties of the two candidates for the highest office. The CDU/CSU polled 44.5 per cent in 1980, against the SPD's 42.9 per cent.

Here the Liberals again occupied a pivotal position in the federal party system. The fate of FJS had really been sealed by the FDP's point-blank refusal to work with him, since it could be assumed that either one of the major parties would require the services of a junior coalition partner in order to form a government. In 1980 there were no other parties, apart from the FDP, able to play such a role at federal level. One of their election slogans actually invited people to vote for the continuation of the Schmidt/Genscher coalition and **against** Strauß.

Although the CDU/CSU result in 1980 under Strauß (44.5 per cent) was marginally higher than that of the SPD, it was below that of the Union parties in 1976 (48.6 per cent) and 1983 (48.8 per cent), both under Kohl as chancellor candidate, despite the criticisms so often levelled at him by "the big Bavarian". There was certainly a school of thought in the CDU at the time which subscribed to the view that Strauß, having been given his chance and having failed, should return to his native Bavaria and stay there.

Following the re-election of Schmidt and the SPD/FDP federal government in 1980, FJS in the first instance did return to Bavaria. Nevertheless he did not stay there. His position as Bavarian minister president enabled him to exert influence in the second chamber in Bonn, the Bundesat. Although he relinquished his seat in the Bundestag in 1978 in order to dedicate his energies to Munich, Strauß still spent time in Bonn during the Eighties. From October 1983 until October 1984 he was president of the Bundesrat and as such represented it in many meetings with foreign heads of state.

After the *Wende* of October 1982/83, it was reported that Kohl offered Strauß every cabinet post in the new governmernt in Bonn – except the two which he would have accepted, namely Foreign Minister or Economics Minister – even that of Defence Minister! By skilfully resisting the strenuous demands by Strauß and Schmidt (this was probably the only thing the two ever agreed on) for immediate new elections, in order to confirm the change in power which had taken place on 1 October 1982 at parliamentary level only – the Federal Republic's first successful constructive vote of no confidence – Kohl was able to delay the federal election until March 1983. That gave the much maligned FDP – accused of treachery and disloyalty (*Verrat und Untreue*) – the time which the party so desperately needed to restore at least some of its badly tarnished reputation.

Had the elections been held immediately after the change in Federal Chancellor from Schmidt to Kohl – made possible only by the fact that 33 of the 52 FDP parliamentary deputies voted for Kohl – the FDP would almost certainly have failed to clear the five-per-cent clause. That would have suited Strauß perfectly of course. The scarred relationship between FJS and the Liberals ever since the Spiegel Affair was again clearly evident for all to see. The CSU, and particularly its leader, wanted to increase their share of cabinet posts and influence in the new conservative-led government at the expense of the FDP. If Strauß, after his 1980 failure, could no longer achieve his life-long ambition of being Federal Chancellor, he still hoped for an opportunity of realising his other ambition of replacing the FDP's Genscher as Foreign Minister.

FJS as "Foreign Minister"

When none of this proved possible, FJS, unpredictable as ever in the eyes of many Germans, began to play the role of a sort of *Ersatz* Foreign Minister. The CSU had always believed in drafting its own foreign policy, and FJS had always believed in visiting foreign heads of state. During this period he visited Mao Tse-tung in China, Pinochet in Chile, Eyadema in Africa and Vorster in South Africa.

In July 1983 FJS arranged a one billion DM bank loan for Honecker's GDR. Given his long-standing and well known contempt for socialist/communist politics and his hard-line opposition to the GDR regime for so many years, his action was greeted with incredulity, both inside and outside his own party. As a direct result FJS received his lowest ever vote (77 per cent) at the biennial elections of the CSU party chairman on 16 July 1983.

Supporters of FJS's actions would argue that he gained an easing of travel restrictions between East and West Germany, as well as other alleviations for the ordinary citizens in the GDR. His visit certainly appeared to be welcomed by the ordinary GDR citizens. He also visited Czechoslovakia and Poland in 1983. Strauß continued to surprise, even shock, people with his unpredictable and individualstic approach, always insisting on setting his own political agenda, including the area of foreign policy.

In June of the following year Strauß's wife, Marianne, died in a car accident, the exact circumstances of which were never fully explained. From 1985 onwards FJS was sometimes accompanied by Renate Piller, a 41 year-old from Salzburg; according to press reports, Frau Piller, who had been applying to an Austrian bishop for an annulment of her first marriage, was viewed with suspicion at the time by Strauß's three grown-up children.[7]

The plans to construct a highly controversial nuclear waste re-processing plant (WAA) in Wackersdorf, near Schwandorf in the Upper Palatinate, had been causing consternation for some time, and the issue was a highly controversial theme in the 1986 Bavarian state elections. When building began in March 1987, some of Bavaria's worst and most violent clashes between police and demonstrators occurred. FJS had always been an enthusiastic proponent of nuclear energy.

---

[7] Reported in a special Extra-Ausgabe of Stern "Das was Franz Josef Strauß" on 7 October 1988, p.78.

In late December 1987 Strauß flew to Moscow with Stoiber, Tandler and Waigel for talks with Gorbatschow. In his memoirs FJS said he drew the conclusion that the new Soviet leader really did want reforms and no more military disputes with the West. Strauß advised neither abject pessimism nor exaggerated optimism vis-a-vis the new developments in the Soviet Union, where he sensed, in 1987, that things were in a state of flux, ready for radical change. FJS recorded in his memoirs that realism, composure and vigilance were required in this regard.[8]

On Saturday 1 October 1988 Strauß, who had returned from his holiday home in southern France only four weeks earlier, was driven to the Oktoberfest on Munich's Theresienwiese. In the afternoon he was flown by helicopter to the woods near Regensburg, owned by Prince Johannes von Thurn und Taxis to go deer-hunting. However, shortly after his arrival Strauß collapsed and was rushed to the "brothers of mercy" hospital in Regensburg. Franz Josef Strauß died, aged 73, on Monday 3 October 1988. No single, precise cause of death was given.

Anyone who saw the amazing scenes of the state funeral procession in Munich and the thousands upon thousands who followed on foot could be left in no doubt whatsoever as to the position Strauß occupied in Bavaria. No King or Queen, no president or Head of State of any country anywhere could have received more attention. As the German press put it, Munich laid on funeral ceremonies which had not been seen since the death of King Ludwig II. The writer Sebastian Haffner commented that, on the day of the funeral procession, Strauß, even more in death than in life, proved that Bavaria was a nation state.

An assessment of the contribution of FJS

A few days after his death a special supplement of the German illustrated magazine Stern appeared. It was dedicated entirely to the life and political career of F.J. Strauß and used the subtitle *"verehrt, verdammt, verkannt"* – revered, condemned, misunderstood. He was described as the man who divided the nation more than any other politician in the Federal Republic, adored by some, feared by others.

---

[8]op.cit. Erinnerugen, pp.564/565

Helmut Schmidt said of Strauß, "this man is dangerous," and Herbert Wehner once called him "an intellectual terrorist." A torrent of adjectives have been used to describe FJS. Three of the adjectives most frequently used by crtitics of Strauß were unpredictable, unscrupulous and uncontrolled. They considered him pugnacious, quarrelsome, explosive, aggressive, insensitive, tactless and impetuous; for many he lacked finesse and judgment when it really mattered.

His supporters, on the other hand, speak of Strauß as having possessed an elephantine memory, a shrewd, analytical mind, and they praise his brilliant oratory, passionate debating skills, superb command of language and rare sense of humour. There were undoubtedly occasions when Strauß could exude charm and wit. FJS's numerous fans saw him as dynamic and decisive, the sort of highly intelligent trouble-shooter Germany needed.

Almost everyone seemed to agree that Strauß was that rare creature, a full-blooded politician (*Vollblutpolitiker* was the most frequently used term in tributes paid after his death) and a political founding father of the postwar Federal Republic, whose colourful personality made many of his contemporaries pale into insignificance by comparison. When opinions were expressed on FJS, few people were neutral. Most were either fervent advocates of the man and his policies or else bitter opponents.

FJS was an original, never boring. He was a person of great contrasts, with the result that on different occasions he sometimes displayed almost opposing characteristics. There was, for example, a sensitive and quiet side to his nature, perhaps not often seen in public. Equally, FJS could be loud and aggressive and seemingly insensitive with his political opponents. He could on the one hand sometimes show real understanding for the everyday problems of ordinary people and relate well to them. For example he revelled in wearing fancy dress for the German carnival season (*Fasching* as it is known in southern Germany) and loved to appear as a man of the people at parties and celebrations like the *Oktoberfest*. On the other hand his advisors often had to persuade him to omit the Latin or Greek phrases and erudite references he so cherished from his writing, on the grounds that most people would simply not understand them.[9]

---

[9] This point, about Straß's love of Latin or Greek quotations, was mentioned to the author in an interview with Wilfried Scharnagl, the editor-in-chief of the CSU weekly Bayernkurier, in Munich on 19 October 1993.

The eminent and respected professor of international politics, Alfred Grosser, concluded that on balance the negative aspects of Strauß's career outweighed the positive ones. He conceded that during the time of the Grand Coalition FJS had been a good Finance Minister, but considered that Strauß's political behaviour was too often unpredictable, for example in the area of *Ostpolitik*, yet remarkably consistent in his dealings with Chile and South Africa.

Many praised the contribution of Strauß in helping to shape the development of both Bavaria and the Federal Republic in the post-war period from the very beginning (*ein Mann der ersten Stunde*). Marion Dönhoff, editor of the reputable German weekly die Zeit, admired his intuition and conviction in relation to the new developments in *glasnost* and *perestroika* in Moscow.

Many of his former opponents, including for example Rudolf Augstein, editor of the weekly news magazine *der Spiegel*, and Willy Brandt, a long-standing political opponent, tried hard to be objective and weighed Strauß's good and bad points after his death. Augstein praised FJS's achievements for Germany. He called him "surely the most significant Bavarian politician since 1918," and showered great praise on his gifts as an orator. Whilst Augstein recognised that Strauß was a gifted and highly intelligent politician, he nonetheless cast serious doubts on FJS's political judgment in several individual cases.

Willy Brandt's remarks when the CSU leader died were magnanimous, given the treatment FJS had meted out to him. Brandt described his erstwhile political "enemy" as "one of the few great talents of German politics." Others, like the Green politician Jutta Ditfurth, were more direct in their open criticism of Strauß the politician: Whilst offering condolences to the immediate family of the Bavarian leader on a personal level, Ditfurth stated that his politics "stood for everything which I detest." Petra Kelly of the Greens was also critical of his "merciless treatment of political opponents."

Karl-Heinz Hiersemann, a former leader of the SPD parliamentary party in Bavaria made the comment in 1988 that FJS was the only Bavarian minister president who never met the leader of the Opposition for a face-to-face talks.[10] Other Bavarian minister presidents did so. This happened

---

[10] Comments reported in op. cit. Stern Extra-Ausgabe 7.10.88, pp.94-97

again in 1994, when Edmund Stoiber invited Renate Schmidt of the Bavarian SPD to the State Chancellory for talks.

The political affairs

The extremely lively and eventful political career of Franz Josef Strauß was overshadowed by a number of political affairs and scandals, in which allegations and rumours abounded, not always backed up by hard evidence and proof. This aspect of his political career dogged Strauß right up to the end and indeed continued even after his death, when new allegations emerged. After an announcement by premier Edmund Stoiber early in 1994, the Strauß children were still fielding questions on behalf of their late father relating to DM 300,000 (around £100,000) annual "extra earnings" (*Nebenverdienst*) from the Friedrich-Baur-Foundation, from which FJS's successor Max Streibl had also benefitted.

In 1993 Strauß's role in the Zwick Affair was still being discussed in the media and his name was mentioned again on 19 March 1994 when Gerold Tandler, one of FJS's former protégés, resigned his post as one of the four CSU deputies to the party chairman over the Zwick Affair.

The only charge ever proved against Strauß, despite his name being linked to several political affairs, was that he lied to the *Bundestag* over the Spiegel Affair. He resigned from his cabinet post as Defence Minister in December 1962 as a result. Since his critics maintained that the name of Strauß was connected with too many dubious episodes in both Bavarian and German politics, and that this prevented him from ever gaining the highest office in the land, brief reference will be made to them.

The HS 30 Affair

The negotiations with the Swiss firm Hispano Suiza over the supply of tanks for the newly established Bundeswehr began under the FRG's first Defence Minister, Theodor Blank. His successor, Strauß, obviously had the opportunity of changing the supplier, but chose not to do so. Between February and August 1957 FJS ordered 10,680 HS 30 tanks/armoured personnel carriers. A total advance of DM 205 million, the highest sum which at that time Bonn had ever paid for a single armaments contract, was paid to Hispano Suiza, although no prototype model had been made available for inspection. It was equally surprising to learn that the firm had never built tanks before.

When the first equipment delivered was shown to have considerable defects, questions were asked in the Bundestag about possible links between a supplier to the Swiss firm and a member of the CDU/CSU executive. Allegations appeared in the German media of bribery and corruption in connection with the huge defence contract. Queries were raised as to why Defence Minister Strauß had approved the investment of so much government money in a firm which had no experience in the relevant field and which in the event proved incapable of delivering satisfactory military equipment. The committee reporting on an investigation of the HS 30 affair in the Sixties reached no clear conclusion. For some the suspicions have always remained.

The Deeg Connection

There was also a lot of speculation during this period concerning Dr Peter Deeg, a so-called *Spezi*, or special friend of FJS. Deeg, a solicitor, had a chequered past[11]. Although Deeg (a member of the CSU) had no experience in the armaments business, he visited Strauß in Bonn representing an Italian grenade manufacturer and managed to arrange a DM 22 million order with the Defence Ministry. The deal reputedly brought him personally around DM 80,000 in commission and fees.

In a completely separate incident in 1957 Deeg acquired a piece of land in Bavaria for DM 450,000; in 1961 he sold it to the federal ministry of defence, which apparently required the land in question, for about three times the price, making a profit of approximately DM 1 million. No concrete proof of any irregularities was forthcoming, but it seemed to some commentators at the time that a number of vital questions had been left unanswered.

---

[11] According to Bernt Engelmann, Deeg was a close friend and colleague of Julius Streicher (NSDAP Gauleiter in Franconia, publisher of the fiercely anti-semitic weekly der Stürmer) during the Third Reich. He joined the CSU in 1945 and, according to Engelmann, became a friend of Josef Müller and Strauß. Bernt Engelmann. Das neue Schwarzbuch. Franz Josef Strauß. Kiepenheuer u. Witsch. Köln. 1980, p.66.

The Starfighter Affair

In October 1958 FJS at the Defence Ministry placed an order for F-104 G (German) Starfighter aircraft for the West German airforce with the American firm Lockheed. Almost at once complaints were made that these interceptor aicraft were being re-constructed to convert them into bombers. Strauß placed an order for 700 jets (starfighters) before any prototype had been tested. The Federal Audit Ofice criticised the acquisition of the aeroplanes more sharply than in any case which had occurred until then.

The safety record of the aircraft was diabolical. A total of 269 starfighters crashed; a total of 110 German pilots were killed. Again serious doubts existed over the precise Lockheed-Strauß relationship, and in particular some financial experts wondered why such an extremely lucrative defence contract had been placed with a manufacturer which had supplied such dubious equipment. How was it possible for the contract to have been continued with such an appalling safety record?

The "Uncle" Aloys Affair

Dr Aloys Brandenstein was an unsuccessful businessman living in very modest circumstances in Frankfurt. He was in debt – to such an extent, apparently, that he had even been barred from his local public house, where he owed money. One year later Brandenstein was a very rich man.

At the end of the war Brandenstein had lived in Bad Aibling in Upper Bavaria. He became a friend of Dr Max Zwicknagel, who owned a brewery. Zwicknagel later became a CSU member of parliament and lived in nearby Rott am Inn. Zwicknagel's daughter, Marianne, who married F.J. Strauß in June 1957, referred to Brandenstein as "Onkel" Aloys. In 1958, down on his luck, Aloys got in touch with his old friend Max Zwicknagel. It was suggested that a visit to Bonn would be in order.

Someone apparently lent Brandenstein the money to travel from Frankfurt to Bonn and visit Marianne Strauß and her husband. "Uncle" Aloys was made most welcome and even talked to Colonel Becker at FJS's Ministry of Defence. A few months later Aloys was acting as a go-between, purchasing items on behalf of the German army from a Belgian weapons factory. He was, with the discreet assistance of Strauß's ministry in Bonn, the advisor for the Remscheid tank manufacturer Erwin Backhaus.

When this firm later got into financial difficulty and had to be sold, Brandenstein quickly found a buyer in Karl Diehl, a Nuremberg industrialist who was a friend of Strauß and a CSU supporter. Business, in the form of regular orders from the Ministry of Defence, took off and was booming in the years that followed. Diehl later found a director for the Remscheid firm from the CSU – Dr Bernhard Goppel, the son of the Bavarian minister president Alfons Goppel, who made way for FJS in 1978.[12]

The FIBAG Affair

FIBAG was the abbreviation for a firm, established on 6 May 1960, called *Finanzbau Aktiengesellschaft*. A joint German/American venture was intended to build some 5,000 dwellings for US troops stationed in Germany. The head of the organisation was Lothar Schloß, an unsuccessful Munich architect. He joined forces with a Passau publisher, Johann Evangelist Kapfinger, who was to receive 25 per cent of the FIBAG capital as free shares, although he had not invested a penny, in return for his "know-how".

Kapfinger was well connected to the CSU, especially to Friedrich Zimmermann, the CSU secretary general, and he was also a so-called FJS-*Spezi*, a good friend of Defence Minister Strauß, who not only gave the project his backing but also sent a letter, personally recommending Schloß to the American Minister of Defence, Thomas S. Gates.[13]

In fact the Rental Guarenteed Housing Project never materialised – no housing was ever built. Neveretheless it developed into a political scandal. Two witnesses testified under oath that Kapfinger, in the negotiations concerning the distribution of the FIBAG shares, said that it was a pity that he had had to give half of his shares to F.J. Strauß. At first Kapfinger denied having said this, but later admitted it was possible he had said something along those lines, but that it was intended ironically. Kapfinger could not be accused of perjury, as he produced medical certificates relating to the time of the negotiations. The case against him was dropped.

---

[12] Information gathered from several souces, including op. cit. Engelmann, p.114 f. and op. cit. Stern, p.81.

[13] In op. cit. Engelmann, p.125/126, the exact content of a recommendation and also a personal letter to Gates, both from Strauß, are reproduced. Engelmann states that a copy of the letter sent to the Pentagon was sent to Kapfinger on 4 August 1960 and signed Dein Franz Josef Strauß.

In the last week of October 1962 the Bundestag was due to debate the FIBAG Affair. An investigative committee was going to try and establish whether or not FJS was a sleeping partner in the DM 300 million housing project. Owing to the Spiegel Affair and FJS's resignation, the other debate never took place. The final remarkable aspect of the case was the disappearance of the FIBAG files in 1963 when they were supposed to be sent by post from Bonn to a court in Nuremberg. In 1975 the files mysteriously turned up in a locker in the main railway station in Mainz.

The BMW Affair

Dr Hartwig Cramer was a former school friend of FJS, who became a solicitor and regularly helped Strauß in legal cases. When the West German ministry of defence granted NATO equipment orders to the firm Triebwerkbau GmbH in Allach, a subsidiary of the automobile manufacturer Bayerische Motorenwerke (BMW), Cramer was asked by Strauß to investigate the firm, in order to check on its economic viability.

Cramer worked out a rehabilitation plan for Triebwerkbau, which was approved by the Bavarian government on 25 March 1959. A few minor alterations in organisation and personell were required – nothing very much. However, one of the changes in personell was that Cramer should join the board of directors at the firm. In 1960 he did so. The view was held by some at the time that this was FJS's way of thanking his former school friend for the great help he had offered over many years in dealing with Strauß's legal matters, particularly allegations and accusations in the press. Hartwig Cramer died in a car accident in 1963.

After investigating a number of aspects of some of the above-mentioned affairs and finding no definite proof against Strauß, the law courts in Munich came to the conclusion in file 180680/64 that, following FJS's period in office in Bonn as federal Defence Minister, "the smell of corruption remained."

The case of the traffic duty policeman

On 28 April 1958 defence minister Strauß, as commander of the armed forces, issued instructions demanding "exemplary behaviour" from all those in positions of responsibility. An incident on the following day, later widely reported in the German press, convinced some people that the minister himself did not always maintain the high standards he demanded from others.

A policeman on traffic duty, called Herr Hahlbohm, was controlling the traffic at the crossing next to the entrance to the Chancellor's Office (*Bundeskanzleramt*) in Bonn. As FJS's car approached, Strauß told his driver, Leohard Kaiser, to ignore the policeman's signals. He did so, turning quickly and crossed in front of a tram, full of passengers, almost causing a collision. On the way out FJS ordered Kaiser to stop next to the traffic policeman. Strauß asked Hahlbohm if he was intending to report the incident and prosecute the driver. When the policeman replied that he was, FJS demanded his name and said, according to newspaper reports, that he (Strauß) would see to it that Hahlbohm would be moved from that crossing.[15]

Strauß wrote first to Dr Degenthoff, the police commissioner in Bonn, asking for Hahlbohm to be reprimanded and removed from his job in Bonn, and then to Dr Dufhues (CDU), the new Innenminister of North Rhine Westphalia, where there had been a change of government. FJS said in his letter to Dufhues that the incident seemed to be part of a number which he had observed during the time of minister president Steinhoff (SPD). Dufhues rejected the complaint, replying that Hahlbohm had acted in accordance with his duty. Strauß then took out a private prosecution against Hahlbohm in a Bonn court, but without success.

"Interference" in the legal process?

In 1960 the Bavarian State sold a piece of land to a Nuremberg estate agent named Hackel. Although the land was valued at around DM 15 million, it was sold at the "preferential" price of DM 900,000. Soon afterwards Hackel made a five-figure donation to the CSU. This aroused suspicion and the public prosecutor's office in Nuremberg started proceedings for an investigation.

The investigation included, in March 1961, searching the CSU party offices in Nuremberg. Schäfer, the CSU chairman of Central Franconia, protested, but to no avail. He turned to Friedrich Zimmermann, the secretary general of the party, but the latter was temporarily suspended from

---

[15] Reported in der Spiegel on 3 September 1958, but also in several other publications. The incident was referred to in op. cit. Engelmann, p.140, and again in der Spiegel on 10 October 1988.

office at the time, because of accusations of perjury relating to the Bavarian Casino Affair (*Spielbankaffäre*). Schäfer therefore approached the new party chairman, Strauß, who at once, from his Bonn ministry, asked Dr Sauter, who was leading the inquiry, to ring "the Defence Minister" in Bonn. When he did not do so, Strauß telephoned him and asked which political party he and the other lawyers involved in the case belonged to.

When, after a long delay, Strauß was eventually forced to answer to parliament over the issue, he assured the *Bundestag* that he had acted in his capacity as CSU party chairman, *not* as a federal minister. Since, however, FJS had specifically requested that Sauter telephone the "federal minister for defence," FJS eventually had to admit to parliament, that as a result of the excitement regarding the investigations, he had committed "a procedural error," which he regretted. Critical voices at the time spoke of unconstitutional attempts to interfere with the courts and the legal process via the abuse of high office for political purposes.

The Barth Case

On 14 September 1961 two jet bombers from the Lechfeld air force squadron flew across the GDR border by mistake and were therefore flying over East German territory. The squadron wing commander was Siegfried Barth. The planes landed at Tegel Airport in Berlin. One day later, after an evening discussion in a bar with air force General Kammhuber, Defence Minister Strauß ordered Barth to be relieved of his duties and employed elsewhere, without summoning the reprimanded wing commander to account for himself.

On 16 September the order was carried out, without any trial or proper examination of the case at all. Barth was moved to the air force office (*Luftwaffenamt*). A promotion which had been planned for 1 October was cancelled. Barth filed a complaint against Strauß. The Bavarian Senate dealt with the matter on 20 December 1961. Four generals and two lieutenants were to appear as witnesses. However FJS's secretary of state Hopf, who was sent to Munich, announced that permission for the witnesses to be heard had been withdrawn.

The Senate discussed the matter and decided that the witnesses should be heard. Hopf insisted that his minister had forbidden the testimonies to be given. The Senate cancelled the hearing, in order to avoid further embarrassment, and on 12 February 1962 met again, reaching a decision on the case: Barth's complaint was upheld and he was to be re-instated. However the re-instatement did not take place.

On 12 May 1963 the affair was reported in the press. The matter was discussed in the *Bundestag*, where Strauß claimed he had behaved quite correctly. Eventually Barth demanded damages for an affront to his reputation and asked that the court's decision to re-instate him be carried out; there was now mounting parliamentary and public pressure. After further delays, Strauß finally obeyed the court order. When the question of financial compensation was mentioned, FJS was reported to have been unwilling to pay himself, but then to have replied that he would pay for everything – if necessary from CSU party funds.

The Spiegel Affair

It must be emphasised that in all the above cases allegations and counter allegations abounded and unsubstantiated rumours were common. Hard facts and concrete evidence were often conspicuous by their absence. Adenauer once said that when someone is pursued by affairs as frequently as Herr Strauß, there must be something in it. However, the only thing ever proved against Franz Josef Strauß was that he admitted having lied to the Bundestag over the Spiegel Affair, one of the most widely reported political events, both at home and abroad, in the history of the Federal Republic. As a direct consequence, he was eventually forced to resign from ministerial office.

On 3 October 1962 the West German news magazine der Spiegel (no. 40/1962) published an article entitled "Balkan in Bonn?" The article asked a series of questions about "Onkel Aloys," Peter Deeg, Kapfinger and FIBAG and the role of Defence Minister Strauß. The article was written by Rudolf Augstein, the editor, signed with a pseudonym Moritz Pfeil. In the previous week's edition of Spiegel (no. 39/1962) the Aloys Brandenstein and Peter Deeg affairs had been discussed in greater detail.

On 10 October Spiegel (no. 41/1962) printed an article on the Bundeswehr and defence policy. It was called *Bedingt abwehrbereit* Fallex 62) – conditionally ready to resist. Fallex was the code name for a NATO military exercise. The article suggested that the Federal Republic's defences were perhaps not as efficient as they might be.

A party colleague of FJS, von der Heydte, reported that an offence had been committed by the publication of the article. It was claimed that the Spiegel article had security implications and amounted to treason, (*Landesverrat und landesverräterische Fälschung*), although some people thought the article by Conrad Ahlers, one of the Spiegel editors, contained nothing that was not already known. Others thought that the real source of the problem was what Spiegel had published about Strauß in previous weeks.

After the Minister of Defence had been consulted and given his views, and after the appropriate federal judge in Karlsruhe had given his signature, search warrants for Spiegel's offices in Hamburg were issued on 23 October, as were orders for the arrest of Rudolf Augstein, the publisher of the news magazine, and Conrad Ahlers, who penned the offending piece. At about 9 pm on 26 October 1962 around eight Bonn security agents, led by federal prosecutor Buback, entered the Spiegel offices. A newspaper cartoon in the Süddeutsche Zeitung on 3/4 November 1962 showed men being marched out of their homes wearing pyjamas (recalling images of the Third Reich).

On 22 October von der Heydte, who had been warmly recommended by FJS for promotion to brigadier general some time before, received his promotion. He explained in an interview with the German magazine Stern that the promotion was not connected to his actions regarding Spiegel – it was not, so-to-speak, "a reward." Defence Minister FJS told a plenary session of the Bundestag that the action against Spiegel was not an act of revenge on his part – he had in the truest sense of the word nothing to do with the matter. ("*Ich habe mit der Sache nichts zu tun, im wahrsten Sinne des Wortes nichts zu tun*".)

This was also the time of the Cuban missile crisis. On the evening before key decisions were to be taken in this connection, there was a reception given by the federal president, Lübke, in Schloß Brühl, attended by many politicians, including Strauß. FJS is alleged to have been less than sober and to have said that during the next few days action was to be taken against Spiegel.

Strauß made a telephone call to Madrid and spoke to the military attaché there, Achim Oster, whom he had known since his Schongau days, and arranged for the author of the offending Spiegel article, Ahlers, who was on holiday in Torremolinos, to be arrested by the Spanish police. This, despite the fact that FJS had apparently been warned by the Justice Ministry that in a case of this nature the involvement of Interpol was not permitted. The federal Justice Minister, Stammberger (FDP), who shortly before had been the only minister who had voted for the continuation of proceedings against Strauß in the FIBAG Affair, was not informed of the action to be taken against Spiegel.

The affair at once provoked strong reactions, even outrage by the general public. There were street demonstrations, student protests and a public outcry over the furtive way in which the whole affair had been carried out "under cover of darkness" (*eine Nacht- und Nebelaktion*), reminiscent of the all-too-frequent midnight arrests under Hitler.

The incident caused tremendous unrest and great concern in West Germany about the freedom of the press, the rights of the individual vis-a-vis the state in a liberal parliamentary democracy, especially in the context of former political regimes in Germany, and a furious public debate ensued about what restrictions should apply to the authority and power of individual ministers. Adenauer denied that he had given his approval for all the events which took place. After a week of questions in parliament in Bonn, Strauß first tried to define what constituted a lie, then stated that he had never said he had nothing to do with the matter (even though he had and his remarks were recorded in parliament).

A new expression entered the German language in the autumn of 1962, comparable perhaps with the new phrase which entered the English language in Great Britain some years ago during the Spycatcher case: "being economical with the truth". The CSU minister Hermann Höcherl suggested that the actions against Spiegel had been "somewhat outside legality" – *etwas außerhalb der Legalität*. Friedrich Zimmermann (CSU) tried to placate friends in his party with the explanation that Strauß had lied for the good of the German people.

When he rang Oster in Madrid in the midst of the affair, Strauß had said that he had just come from the Federal Chancellor and that what he had to say was an order from him. Adenauer denied this. Nonetheless the order was followed at the time and Spiegel editor Ahlers, and his wife, were taken into custody. FJS later referred of his role in the Spiegel Affair as being essentially that of an assistant telephonist.

FJS did not resign voluntarily. The FDP federal ministers in Bonn threatened to resign, and thus bring down the government, if Strauß refused to stand down as Defence Minister. Eventually he resigned on 11 December 1962. In his memoirs FJS said of the Spiegel Affair that truth had had no part to play in the way the matter was handled; emotional reactions and fuelled-up feelings had dominated the proceedings. He recorded that he felt he had been treated "like a Jew who had dared to appear at the Reich Party Conference of the NSDAP."[16]

Such an episode would have spelt the end of many a political career, but not in the case of Strauß, who not only survived and carried on almost as if nothing had happened, but even returned to Bonn with a ministerial portfolio four years later. Despite this, however, his career was badly tainted, as was the relationship between Strauß and the FDP.

These affairs even brought Strauß's judgment into question in the final phase of his life. His decision to exempt private pilots, of which he himself was one, from the increase in aircraft fuel in 1987 was very strange, to say the least. This was probably more a sign that the Bavarian leader was losing his touch, rather than the start of another affair. As usual, however, there were those who supported his action, saying the ruling applied to only a very small proportion of private pilots but was a significant and justifiable change for large airlines.

However the Zwick Affair even followed FJS to his grave and beyond. Strauß was a friend of Eduard Zwick, the "swimming-pool king" from Lower Bavaria, who was granted a massive loan by the Bavarian state. Zwick lived in Switzerland for many years and was one of Germany's greatest tax debtors. He still owes over DM 70 million to the Bavarian state, and in 1995 was still living in exile in Switzerland in order to avoid repaying the tax he owed. Gerold Tandler, a former Bavarian Finance Minister under FJS, once took a personal loan from Zwick and was at one time Zwick's business partner in Altötting.

Tandler became Secretary General of the CSU in 1971, held the offices of minister of the interior, economics and finance in various

---

[16] op. cit. Erinnerungen, p.424.

Bavarian cabinets and it appeared that he was being groomed as a likely "crown prince" by FJS. When he resigned his position as deputy party chairman on 17 March 1994, Tandler was the third CSU politician, following Streibl and Gauweiler, who had to step down from office in the post-Strauß era.

FJS and the CDU/CSU relationship

The fact that Franz Josef Strauß was both a Bavarian and a national politician of great prominence exerted a significant influence on the relationship between the two Union parties and the joint parliamentary party (*Fraktionsgemeinschaft*) which they have always formed at federal level in the Bundestag. The unique role played by the CSU in the federal party system, acting as an autonomous regional party in Bavaria and simultaneously as a federal party (often as a government coalition party) in Bonn, has meant that relations between the two partners have had their ups and downs. This was especially the case during the period 1961-88, when FJS was chairman of the CSU. From 1978 until 1988, when he was minister president of Bavaria as well, the Kohl-Strauß relationship added an extra, often difficult, dimension to the problem.

Padgett and Burkett in 1986 referred to the strong influence of Strauß on CDU/CSU relations, especially during the years 1969-80, in the post-Adenauer period. This occurred at a time when the CSU began to assert its independence more and more. This called into question the automatic dominance which the CDU had hitherto displayed in the electoral alliance.[17] It was in sharp contrast to the very amicable relations between the two Christian Union parties during the first twenty years of political partnership in the FRG, when there was a more cohesive sense of common purpose. This close relationship was further cemented by the personal authority of a domineering Chancellor.

---

[17] Stephen Padgett and Tony Burkett. Political Parties and Elections in West Germany. Hurst. London. 1986, p.128/129.

Influence of the FDP

Another important factor was the developing role of the FDP as a government coalition partner. In the nascent phase of the West German party system the Free Democrats had not yet established themselves in what was initially a multi-party system. Even when the Liberals began to emerge as the third party, they received a setback in 1956 when they resigned from federal government, and an even bigger shock when in 1957 the CDU/CSU became the only party ever to gain an absolute majority of the votes at a federal election. Just as it was starting to carve out a niche for itself, the FDP became superfluous. The 1957 result clearly brought the Union parties even closer together.

The 1969 change in power highlighted the disproportionate and powerful position of the Free Democrats in the Bonn system. The fact that they were seen to be no longer tied to the apron strings as a type of appendix or *Anhängsel* of the CDU/CSU put a new perspective on things. The generally amicable relations which obtained between the FDP and the CDU in some Länder, even during the social-liberal coalition, contrasted sharply with the gradually worsening relations between the FDP and the CSU mainly owing to the Spiegel Affair and a feeling of mutual distrust between Strauß and the liberal party.

In this way "the Strauß factor" and its wider implications had a fundamental effect on the relationship between the two Christian Union parties. Under the leadership of Strauß, the CSU increasingly underlined its role as an autonomous *Landespartei*. One of the ways it did this was by FJS threatening to disband the joint parliamentary arrangements in 1976.

1976: The Wildbad Kreuth split

Before the federal election on 3 October 1976 several disagreements had broken out in the Union parties. Harsh criticism of the parties' joint Chancellor candidate, Helmut Kohl had come from Strauß. On 19 November 1976, at the CSU's private meeting at its party retreat in Wildbad Kreuth in Upper Bavaria, the party voted by 30 to 18 in a secret ballot in favour of terminating the long-standing agreement on forming a joint parliamentary party in the *Bundestag*.

This of course meant a split in the conservative camp; the CSU would campaign throughout the Republic and the CDU would be free to campaign in Bavaria. Suggestions to re-think the decision came first from the Franconian and Swabian CSU associations. The more traditional bastions of support in Old Bavaria, as usual, adopted a hard-line attitude. However Strauß had probably not reckoned with the equally hard-line approach adopted by Kohl and the CDU. On 12 December, two days before the new *Bundestag* was due to convene, the CDU/CSU joint agreement was renewed.

FJS, in his memoirs, expressed the fear that the CSU might fall below the fifty per cent mark after his "reign." He would have certainly been shocked if he had known that in April 1994 the CSU itself would be expressing doubts about its chances of clearing the five per cent hurdle (requiring a Bavarian vote of 37-39 per cent) for the European elections. Under FJS the CSU achieved over 62 per cent in 1979 and over 57 per cent in 1984 at the European elections.

Opinion polls predicted around 40-42 per cent for the CSU in spring 1994, as the Strauß-Zwick revelations in the so-called Bavarian Amigo affairs reached a peak. The deposed Max Streibl invited Franz Schönhuber to his home, at a time when the CSU was trying hard to distance itself from the Republican Party, who were under close observation by the Bavarian authorities. Several sources, including the respected German weekly die Zeit, expressed the firm conviction that Bavaria might well get its first coalition government for over thirty years following the September 1994 state elections.[18] In fact, as will be seen in the following section, the CSU managed to make a dramatic recovery; at the time, however, it was by no means a foregone conclusion that 1994 would see the party return to power in both Bonn and Munich.

There can be no doubt whatsoever that Franz Josef Strauß played an absolutely crucial role in the CSU success story. Whether admired or condemned, Strauß, unpredictable and controversial to the end, was Bavaria's most dynamic and gifted postwar politician. Just like "his" party, Strauß was simultaneously both a national and regional force. His influence – for

---

[18] "Das anscheinend eherne System Bayerns erlebt seine überfällige Säkularisierung. An ihrem Ende könnte die erste Koalitionsregierung seit über dreißig Jahren stehen. Die Zeit. 11 March 1994, p.4.

better or for worse, according to political preference – will be felt in Bavaria for many years to come.

10

Bavarian Politics in the mid-Nineties

Although the Bavarian state election result in October 1990 of 54.9 per cent under Max Streibl appeared impressive, only two years after the death of Strauß, it was widely felt that it had been achieved on the back of the euphoric mood of German Unity, almost regardless of who was leading the party.

The situation again looked critical for the CSU when Streibl had to be replaced by Stoiber in June 1993. The Amigo Affair had broken, and in any case it was felt that Streibl did not really have the kind of personality and leadership qualities to take the CSU into "super election year." With the approach of three key elections for Bavaria, the opinion polls of March/April 1994 forecast the possibility of the first Bavarian coalition government in Munich for nearly thirty years. Although, with hindsight, such a scenario might sound far-fetched, in the spring of 1994 both German polling institute questionnaires[1] and CSU internal party polls[2] showed the

---

[1] The German magazine Focus (15/94, p.19) reported the results of a survey conducted by Basisresearch at the end of March 1994, showing a drop in CSU support from 47 to 42 per cent for the Munich parliament, and down to only 40.5 per cent in Bonn. The same poll predicted a rise in support for the SPD, Greens and FDP at that time.

[2] In an interview with Prof. Heinrich Oberreuter in the Akademie für politische Bildung in Tutzing, Bavaria in April 1994 the author was told that the CSU's own internal research had come up with some predictions which were even marginally below forty per cent.

CSU to be at an all-time low. The clear indication was that in the autumn elections in Munich and Bonn a result of around the forty per cent mark was on the cards.

It should not be forgotten that, at that particular time in Bavaria, the Zwick Affair was receiving maximum publicity and CSU politician and former Strauß-*Spezi* Gerold Tandler had just resigned from the party heirarchy on 17 March, owing to some ambiguous financial dealings which had cast a cloud over his position for some time. There was also a slump in the standing of the Union parties in Bonn generally, accompanied by suggestions that the federal SPD, led by a new chancellor candidate, Rudolf Scharping, might even lead a red-green government in Bonn after the October 1994 federal election.

The 1994 European election – the first test

An early test of public opinion during "super election year" (1994) in all the federal states came in June with the European elections. The depressing predictions for the CSU earlier in the year had led to speculation about whether the Bavarian sister party would clear five per cent of the overall vote in Germany, in order to enter the European parliament. In the event the CSU polled 6.8 per cent, to the great relief of all those who had been working to that end in the party headquarters (*Franz Josef Strauß-Haus*) in the Nymphenburger Straße in Munich.

The CSU in fact gained almost forty-nine per cent of the vote in Bavaria in the European elections. Whilst this was obviously not the absolute majority the party leadership would have liked, it did represent a remarkable recovery in only just over two months. Any figure approaching the fifty per cent mark in a state election is likely to produce an absolute majority of the parliamentary seats.

This was indeed the start of the CSU recovery. The electorate had clearly identified Bavarian interests in Europe with the traditional "party of state,"[3] and the result actually represented an increase of 3.5 per cent on the

---

[3] In a television interview on 13 June 1994, the day after the European election, Theo Waigel, the party chairman said: " . . . die Vertretung der bayerischen Interessen ist mit der CSU identifiziert." This was shown on the Bayerischer Rundschau news programme on Bavarian television.

CSU's – admittedly poor – achievement (45.4 per cent) at the previous European election in 1989. The scene appeared to be set for Edmund Stoiber to set his sights on clearing the much treasured fifty per cent hurdle in the forthcoming Bavarian state elections later in the year.[4]

At the European poll in June the SPD slipped to 23.7 per cent; so Bavaria's main opposition party had not benefitted from any drop in popularity of its rival. Neither had the Liberals, who were going through an extremely bad patch in Germany in general, and in Bavaria in particular. Whilst the FDP failed to clear the five per cent barrier, the Republicans and the Greens had no such problems. Manfred Brunner's break-away party – he is a former leader of the Bavarian FDP – the Association for Free Citizens (*Bund der freien Bürger*) had only very limited appeal. All the other tiny parties which contested the election managed almost seven per cent between them.

Table 10.1

Results of the June 1994 European Elections in Bavaria

CSU	48.9%
SPD	23.7%
Alliance '90/Greens	8.7%
FDP	3.3%
Republicans	6.6%
Free citizens	1.8%
Others	6.9%

Some observers at the time saw the resurgence of the Republican Party in Bavaria as an unwelcome trend, given the falling electoral support Schönhuber's party had been gaining in the state elections which had occurred earlier in the year. Although the CSU result was encouraging, from their point of view, and was obviously a step in the right direction for

---

[4] The same programme showed an interview with Edmund Stoiber, the Bavarian minister president, who said: " . . . ich glaube, daß dieses Wahlergebnis mich darin bestätigt, daß dies (gaining over 50 per cent of the votes and mandates for the CSU in Bavaria) ein realistisches Ziel ist, daß wir erreichen werden und können."

a party which desperately longed to return to its former glory, there was still sufficient doubt as to whether the Republicans were staging a recovery too, based on support transferred from disillusioned CSU voters.

Another indicator, which may have escaped the attention of many observers outside Bavaria, was the result of the elections to the city council (*Stadtrat*) in Munich, also held on 12 June 1994. Although the two major parties were neck-and-neck, the SPD – traditionally very strong in the Bavarian capital – dropped nearly six percentage points compared to the 1990 election (42.0 down to 34.4); the CSU gained over five percentage points (30.1 in 1990, up to 35.5). This was another indication, albeit very much a localised, Bavarian one, of the beginning of a CSU recovery.

The results of the European elections in Germany overall actually produced an ambivalent picture, despite the fact that Helmut Kohl and the CDU were credited in the German media with a resounding victory. In fact the SPD recorded slightly more votes than the CDU, but, as the CDU/CSU votes are always taken together, the joint result looked more respectable, at almost forty per cent – not a particularly impressive joint figure for two of the government parties. It certainly appeared that the CSU recovery was occurring more quickly than that of its sister party.

Table 10.2

Results of the June 1994 European Elections in Germany	
CDU	32.0 %
CSU	6.8 %
SPD	32.2 %
FDP	4.1 %
Alliance '90/Greens	10.1 %
PDS	4.7 %
REPS	3.9 %

Despite clearing the five per cent clause in Bavaria, the Republican Party failed to do so in Germany overall, as did the Free Democrats. In the case of the FDP the European election was the third one in a row, starting with the Hamburg state election in September 1993, at which the German liberal party had fallen below five per cent. Worse was to come The Party of Democratic Socialism – PDS – (*die Partei des demokratischen Sozialismus*) just missed the five per cent hurdle too, but the steady improvement in the Green vote in Germany was refected in their result of over ten per cent.

The 1994 Bavarian State Election – the real test

Between June and the end of September 1994 it was clear that the general situation was improving for the CSU. Nevertheless, it was by no means certain that the party would achieve its avowed aim of gaining over half the parliamentary seats in Munich, and if at all possible, an absolute majority of the votes too. Of the fifteen parties contesting the elections, only seven gained more than one per cent of the valid votes cast, and only three parties entered parliament by clearing the five per cent clause. The right-wing ecology party, the ÖDP managed over two per cent, and the Bavarian Party (BP) one per cent. It was interesting to note that on 12 August 1994, only just over one month before the Bavarian state elections, Rudolf Drasch, the honorary president of the Bavarian Party, left the party, after twenty-five years' membership. It was reported in the Bavarian press that Drasch no longer saw any hope for his party.[5]

Although the CSU lost both votes and seats compared with the 1990 state elections, it did record over fifty per cent of both the votes **and** the seats in parliament – 120 out of 204 – polling over six million of the 11.67 million valid votes (in Bavaria each voter has two votes). Despite the return to the throne for the Bavarian "monarchist" party, the SPD also improved its electoral performance. The Bavarian electorate appeared to have taken a collective decision to support the established people's parties (*Volksparteien*) rather than any protest parties.

They not only confirmed what minister president Stoiber and party chairman Waigel had been hoping for, they also increased the vote for the *königlich-bayerische* SPD, under the seemingly popular leadership of Renate Schmidt, to over thirty per cent for the first time since 1982. The long-awaited answer to the question of whether the Republicans would clear the five

---

[5] "Ich sehe keine Hoffnung für die Partei, nochmals eine ernstzunehmende politische Kraft zu werden," reported on p.3 of the local newspaper in Grafenau on 13.8.94.

per cent hurdle on their "home territory" (*Stammland Bayern*), only three weeks before the federal election, was negative. The Liberals recorded yet another defeat – their worst ever postwar figure in Bavaria. The Bavarian Greens, by way of contrast, fared better; they were the third party to enter the Bavarian *Landtag* in 1994.

Table 10.3

The results of the Bavarian state elections in September 1994

	% of the vote	parliamentary seats
CSU	52.8	120
SPD	30.1	70
Greens	6.1	14
FDP	2.8	
REP	3.9	
ÖDP	2.1	
BP	1.0	
BfB	0.4	
Total		204

The 1994 Federal Election – the final test

Having returned victoriously to power on its home ground, one final challenge remained for the CSU. On 16 October 1994 Helmut Kohl was returned as Federal Chancellor, though with a much reduced majority for his CDU/CSU/FDP coalition government of only ten mandates. His party gained twelve extra seats (*Überhangmandate*)[6] – the highest number in the history of the FRG so far. This actually produced the largest number of German members of parliament in the Bundestag since 1949. The total for the new Germany, normally 656 – 328 elected by the first votes and an equal number elected by the second votes – increased in 1994, owing to the extra mandates,

---

[6] See chapter four. These additional mandates are created when a German party gains more direct, or first-vote, seats in the constituencies than it is, strictly speaking, entitled to, on the basis of proportional representation, according to the second, or party-list votes. This often occurs when one of the two main parties wins either all, or nearly all of the constituencies in a federal state.

to 672. Since both the salaries, as well as the allowances and benefits, offered to German members of parliament are very generous, every German parliament costs the state a great deal of money; the current *Bundestag*, however, is not only the largest but also the most expensive one ever.

These additional parliamentary seats occurred in four of the five new federal states in the east, plus Baden-Württemberg. The SPD also gained four such seats, one in the tiny city state of Hamburg, and three in Brandenburg. Whilst the CDU lost ground compared to the first all-German elections in 1990, the SPD, under Rudolf Scharping, improved its performance.

This SPD improvement was not quite enough to bring about a change in power in Bonn, however, given the fact that the Liberals had already stated their preference for the continuation of a coalition with the Union parties. The crucial question was whether they would poll over five per cent at federal level. In the event they did, despite a series of abysmal performances at regional level. The FPD fell below the five per cent hurdle at ten consecutive elections up to, and including 16 October. The Free Democrats even dropped from 6.5 to 3.8 per cent in the local elections in North-Rhine Westphalia, held on the same day as the federal election, although the five per cent clause does not apply to local elections.

In fact Kohl had a lot to thank the CSU for. Bavaria was the only Land where any party cleared fifty per cent at the 1994 federal elections. The CDU was happy to clear forty per cent in several Länder. The CSU poll of 7.3 per cent in Germany overall improved on the figure of 6.8 per cent at the European elections. It regained its position from the FDP of being the second strongest government coalition partner; at present the CSU is essential for Kohl to be able to govern in Bonn. That was not the case four years earlier, when, in theory, the CDU and the FDP could have ruled alone between 1990 and 1994.

In October 1994 the SPD won only one of the forty-five Bavarian constituencies. Again the voters in Bavaria had apparently decided that the CSU was undoubtedly the party best suited to represent Bavarian interests. CSU hegemony was re-established. The SPD out-voted the CDU in only six of the sixteen federal states at the German federal election in 1994: Hamburg, Bremen, North-Rhine Westphalia, Saarland, Berlin and Brandenburg. Although the Bavarian SPD gained a little ground, in keeping with the national trend, it was not able to break through what for them had become the "magic" thirty per cent barrier, as it had done three weeks before at the Bavarian parliamentary elections.

As usual, the Liberals cleared the five per cent hurdle, even in Bavaria, in the federal election. Extensive research on voting behaviour in Germany reveals that the reason for this is the desire amongst some of the German electorate to give their second vote to the FDP at a federal election, because of its role and specific functions in the German party system, rather than what the party actually stands for. In doing so, those who do support the FDP are casting a vote for a coalition government in Bonn.

This explains why virtually the same Bavarian voters who, at the *Land* election on 25 September 1994 gave the liberal party only 2.8 per cent (the Bavarian FDP's lowest ever result since it was founded in 1946), rewarded the Liberals, only three weeks later on 16 October 1994 at the federal poll, with 6.4 per cent (in Bavaria). This also underlines the fundamental differences between the Bavarian and federal party systems. The pivotal position adopted for so many years by the Free Democrats in the German party system has never been available to them in Bavaria, mainly as a result of CSU hegemony and the very different prevailing circumstances there.

Table 10.4

The results of the 1994 federal elections in Bavaria	
CSU	51.2
SPD	29.6
FDP	6.4
Alliance '90/Greens	6.3
PDS	0.6
REPS	2.8
Others	2.5

A new dimension in the German party system

The Party of Democratic Socialism (PDS) did not contest the 1994 Bavarian state election, since that was not a region where any substantial support could be expected for the successor to the SED, the former East German

Socialist Unity Party (*Sozialistische Einheitspartei Deutschlands*). At the federal elections the PDS gained negligible support in Bavaria (less than one per cent – see above). So that was just a recent example of another clear difference between the German and the Bavarian party systems.

Although the PDS had fought the first all-German federal elections in 1990, it was widely assumed at the time that it was only the special amendment to the electoral law – for that election only a party needed to clear the five per cent hurdle in **either** the western or the eastern *Länder* – which had allowed it to enter the *Bundestag*[7]. In fact the predictions were correct, in the sense that the PDS did not attain five per cent in 1994. Nevertheless many observers probably underestimated the way in which it became a party with specifically east German appeal, gaining 17.6 per cent there, but only 0.9 per cent in the west.

Table 10.5

The results of the 1994 federal elections in Germany				
	%	Seats	West	East
CDU/CSU	41.5	294		
CDU	34.2	244	33.0	38.5
CSU	7.3	50	9.2	
SPD	36.4	252	37.6	31.9
FDP	6.9	47	7.7	4.0
Alliance '90/Greens	7.3	49	7.8	5.3
REPS	1.9		2.0	1.4
PDS	4.4	30	0.9	17.6
Total		672		

In 1994 the PDS was able to benefit from a stipulation in German federal electoral law which suspends the application of the five per cent clause in the case of any party which wins at least three constituencies. The PDS won

---

[7] Several articles on the 1990 federal elections, including one by the author, Germany United: the 1990 all-German elections in West European Politics, Vol. 14 no. 3, suggested that it was highly unlikely that the PDS would clear the five per cent clause (for the whole of Germany) at the 1994 federal election. In fact the PDS, in 1994, polled 4.4 per cent, but because it won four direct constituency seats, it received thirty seats in the Bundestag.

four such "direct" seats, on the basis of the voters' first votes in the federal elction, all in east Berlian. In the light of Bavaria's independent stance, idio-synchratic traditions and special political culture, this new dimension to the German party system was not relevant in Germany's largest federal state. As usual, Bavaria was an exception to the rule.

The future for Bavaria in a new era

The unification of Germany in 1990 clearly created a new political situation. As Stephen Padgett points out, "none of the parties escaped its effects."[8] Both the CSU and the Free State of Bavaria, the largest of the ten western federal states in the old FRG, not only had a lot to lose from German Unity, both the party and the *Land* had already suffered greatly from the loss, in 1988, of F. J. Strauß, their greatest postwar leader. This double blow certain-ly had an initial negative effect on Bavaria, as has been demonstrated.

By 1995, however, it looked as if things were beginning to return to normal in Bavaria. Yet "normal", in Bavaria, often means being an excep-tion to the rule. Any statement, even if it is a generalisation – and all gener-alisations are of course notoriously dangerous – which may have an ele-ment of truth about it with reference to Germany as a whole, will often not be valid for Bavaria, especially where the political system is concerned.

Bavaria in the mid-Nineties was starting to show a return to established political patterns. At its forty-fifth party conference in 1995, encouraged by its improved performance at the autumn 1994 elections, the Bavarian SPD was planning how best to attack the CSU on areas such as rights of asylum. The party delegates voted for the twin leadership team of Renate Schmidt, re-elected as chair of the Bavarian SPD, and Albert Schmid as *Generalsekretär*. The decisions taken at the party conference were immediate-ly criticised by the CSU's new, somewhat controversial party manager, Bernd Protzner, as wrong-headed. Things were definitely back to normal.

The Bavarian Greens, at their 1995 party conference, criticised the SPD, stating that they had no intention of joining forces with a party which refused to launch a serious attack on the CSU regarding the issues that really

---

[8] Stephen Padgett (ed.) Parties and Party Systems in the New Germany. Dartmouth. 1993, p.xix.

mattered most to the B'90/Grüne in Bavaria. These issues included the management of the state finances. So here too it appeared to be "business as usual," in the sense that, once again, the opposition parties in Bavaria were apparently not going to join forces, even though it might be in their best interests to do so, in a party system which is again dominated by one party, holding fifty-nine per cent of the seats in the Munich parliament.

In Spring 1995 there were lively discussions in the Bavarian parliament on proposals to increase the remuneration of Bavarian MPs from DM 8,700 to over DM 11,000 per month. The complaint of the Bavarian Green deputies that a twenty-seven per cent increase was excessive were countered by some CSU parliamentarians, who pointed out that there had been no increase over the previous three years (*eine Nullrunde*).

Meanwhile the "party of state", having returned to power after several scares, is very definitely back in its stride and is now behaving with renewed confidence. In March 1995, for example, the Bavarian government welcomed the President of Ecuador, Duran Ballen, who was on a state visit to the Federal Republic. Stoiber and Waigel did of course not miss a chance to welcome Ballen to Munich, and economic and business agreements were signed with Siemens and other concerns[9]. Edmund Stoiber went on a two-week state visit to China during the first half of April 1995, creating the impression, just as FJS used to do, that Bavaria sees itself as an independent nation state. The CSU has definitely returned to power in Munich with a vengeance. In Bonn, where Kohl's government has, by German standards, a narrow majority and also a weakened coalition partner in the FDP, Waigel's party may still have some battles to fight.

The CSU is a unique party which has played a unique role in both the German and Bavarian party systems during the last fifty years, since its founding in 1945/46. This fact, combined with the idiosynchratic, independent-minded development and distinct political culture of the Free State of Bavaria, means that, despite the ups-and-downs of recent developments, Bavaria still undoubtedly constitutes a special case – indeed a

---

[9] On 18/19 March 1995 the President of Ecuador, Duran Ballen, had discussions with the Bavarian government on a range of business contracts, as well as visiting Schloß Linderhof. Such "state visits" have always taken place in Bavaria, which still adopts the mentality of considering itself to be a separate unit, far more important than the other federal states in Germany.

remarkable phenomenon, worthy of special study, amongst the present-day German *Länder*.

It seemed as if the wheel had come full circle in March 1995 when the name of the Strauß family again hit the media headlines in a negative context. The Bavarian minister president publicly criticised his state secretary in the education ministry, Monika Hohlmeier, FJS's daughter, who had organised the purchase of some educational equipment for Bavarian Schools. Stoiber accused Hohlmeier of "behaving incorrectly," when it came to light that her brother, Max Strauß, a thirty-five year old lawyer, had been involved in promoting the firms involved in the purchase. Stoiber said that the incident was "out of order." Two days later the Bavarian education minister, Hans Zehetmair, also rebuked Frau Hohlmeier for her unwise actions, which he said might have led to "a conflict of interests."[10] As a result of the incident, Stoiber withdrew the field of multi-media from Frau Hohlmeier's area of responsibility. In autumn 1994 she lost responsibility for her specialist area of kindergarten provision, because Stoiber had lost confidence in her ability.

Edmund Stoiber already had reason to believe that the Strauß children were becoming an irritation whilst he was trying to project himself as "Mr Clean" (*Saubermann*). Stoiber made great play of the fact that he had not accepted any money from the Friederich Baur Foundation as executor (*Testamentsvollstrecker*), although his two predecessors (Max Streibl and FJS) had done so – somewhat dubiously. When Stoiber suggested that Strauß's heirs repay at least part of the DM one million, they refused. Indeed Max Strauß said at the time that anyone attempting to recover any of the money would get a "bloody nose." Normality has certainly returned to Bavarian politics, as long as you remember that things are different there: *in Bayern gehen die Uhren anders*.

---

[10] Spiegel 13/1995, p.28 article "Blutige Nase".

Appendix 1

State elections in Bavaria 1946 to 1994

| Year | Turn-out in % | Share of the vote in % | | | | |
		CSU	SPD	FDP	Greens	Other
1946	75.7	52.3	28.6	5.7	–	13.5
1950	79.9	27.4	28.0	7.1	–	37.5
1954	82.4	38.0	28.1	7.2	–	26.7
1958	76.6	45.6	30.8	5.6	–	18.0
1962	76.5	47.5	35.3	5.9	–	11.3
1966	80.6	48.1	35.8	5.1	–	11.0
1970	79.5	56.4	33.3	5.6	–	4.7
1974	77.7	62.1	30.2	5.2	–	2.5
1978	76.6	59.1	31.4	6.2	1.8	1.5
1982	78.0	58.3	31.9	3.5	4.6	1.7
1986	70.1	55.8	27.5	3.8	7.5	5.4
1990	65.9	54.9	26.0	5.2	6.4	7.5
1994	67.9	52.8	30.1	2.8	6.1	8.2

Appendix 2

Federal elections in Bavaria 1949 to 1994

Year	Turn-out in %	Share of the vote in % CSU	SPD	FDP	Greens	Other
1949	81.1	29.2	22.7	8.5	–	39.5
1953	86.0	47.8	23.3	6.2	–	22.6
1957	87.7	57.2	26.4	4.6	–	11.8
1961	87.2	54.9	30.1	8.7	–	6.2
1965	85.9	55.6	33.1	7.3	–	3.9
1969	85.2	54.4	34.6	4.1	–	7.0
1972	89.8	55.1	37.8	6.1	–	1.0
1976	89.6	60.0	32.8	6.2	–	1.0
1980	87.6	57.6	32.7	7.8	1.3	0.5
1983	87.6	59.5	28.9	6.2	4.7	0.7
1987	81.7	55.1	27.0	8.1	7.7	7.7
1990	74.4	51.9	26.7	8.7	4.6	3.1
1994	77.0	51.2	29.6	6.4	6.3	2.5

Appendix 3

European elections in Bavaria 1979 to 1994

Year	Turn-out in %	Share of the vote in %				
		CSU	SPD	FDP	Greens	Other
1979	58.9	62.5	29.2	4.7	2.9	0.8
1984	46.2	57.2	27.6	4.0	6.8	4.5
1989	61.1	45.4	24.2	4.0	7.8	4.0
1994	56.5	48.9	23.7	3.3	8.7	8.7

* In 1989 the Republicans gained 14.6%

In 1994 the Republicans gained 6.6%

References

Chapter One

Bulmer, S. (ed.) (1989) The Changing Agenda of West German Public Policy. Dartmouth. Aldershot.

Burkett, T. and Padgett, S. (1986) Political Parties and Elections in West Germany. Hurst. London.

Childs, D. (1991) Germany in the Twentieth Century. Batsford. London.

Dalton, R. (ed.) (1993) The New Germany Votes. Berg. Oxford.

Hübner, E. and Oberreuter, H. (1992) Parteien in Deutschland zwischen Kontinuität und Wandel. Bayerische Landeszentrale für politische Bildung. Munich.

Kolinsky, E. (1992) Inaugural Lecture at the University of Keele, published as People and Politics in the Unified Germanies; A Citizens' Germany in Europe?

Metzel Poeschel (1993) Das Statistische Jahrbuch für die BRD. Wiesbaden.

Paterson, W. and Southern, D. (1991) Governing Germany. Blackwell. Oxford.

Mintzel, A. and Oberreuter, H. (1992) Parteien in der BRD. Leske und Budrich. Opladen.

Rohe, K. (1990) Elections, Parties and Political Traditions. Berg. Oxford.

Smith, G. Paterson, W., and Merkl, P. (eds.) (1989) Developments in West German Politics. Macmillan. Basingstoke.

Stöss, R. (1990) Parteienkritik und Parteienverdrossenheit. Aus Politik und Zeitgeschichte. Beilage zur Wochenzeitung das Parlament. B 21/90. Bonn.

Strauß, F. (1989) Die Erinnerungen. Siedler. Berlin.

Chapter Two

Aiblinger, S. (1975) Vom echten bayerischen Leben. BLV Verlagsgesellschaft. Munich.

Bäuerlein, H. (1970) Die Bayern in Bonn. Seewald. Stuttgart.

Bavarian State Chancellery (1991) Information über Bayern. Munich.

Bayerische Landeszentrale für politische Bildungsarbeit brochure (1989) Freistaat Bayern. Eine kleine politische Landeskunde. Fourth edition. Munich.

Beier, B. et al (1983) Die Chronik der Deutschen. Chronik Verlag. Harenberg. Dortmund.

Bosl, K. (1965) Zur Geschichte der Bayern. Wissenschaftliche Buchgesellschaft. Darmstadt.

Bosl, K. (1971) Bayerische Geschichte. Territorienverlag. Munich.

Bosl, K. (1980) Bayerische Geschichte. dtv. Second edition. Munich.

Dollinger, H. (1976) Bayern. 2000 Jahre in Bildern und Dokumenten. Bertelsmann. Munich.

Fest, W. (1978) A Dictionary of German History. 1806-1945. Proir. London.

Fink, A. (1965) Bayern in Europa. Buchner. Munich. 1965.

Gebhardt, J. (1982) Bayern. Deutschlands eigenwilliger Freistaat.. Historisch-gesellschaftliche Aspekte der politischen Kultur in Bayern. Bayerische Landeszentrale für politische Bildunsarbeit. Munich.

Golay, J. (1965) The Founding of the FRG. University of Chicago Press. Second impression. Chicago.

Habisreutinger, J. and Krick, W. (1959) Geschichte der neuesten Zeit. 1815-1950. Bucher Verlag. Third edition. Bamberg.

Hamrecht, R. (1976) Der Aufstieg der NSDAP in Mitell- und Oberfranken (1925-1933). Schriftenreihe des Stadtarchivs Nürnberg. Band 17. Erlangen.

Hoffmann, H. (1985) Bayern. Handbuch zur staatspolitischen Landeskunde. Olzog. Eighth edition. Munich.

Hubensteiner, B. (1977) Bayerische Geschichte. Staat und Volk, Kunst und Kultur. Süddeutscher Verlag. Munich.

Kershaw, J. (1983) Popular Opinion and Political Dissent in the Third Reich. Bavaria 1933-45. Clarendon Press. Oxford.

Milatz, A. (1968) Wähler und Wahlen in der Weimarer Republik. Schriftenreihe der Bundeszentrale für politische Bildung. Heft 66. Second edition. Bonn.

Nöhbauer, H. (1987) Chronik Bayerns. Chronik Verlag. Dortmund.

Rall, H. (1974) Zeittafel zur Geschichte Bayerns. Süddeutscher Verlag. Munich.

Roth, R. (1993) Freistaat Bayern. Politische Landeskunde. Bayerische Landeszentrale für politische Bildungsarbeit. First edition 1992. Second edition 1993. Munich.

Rübesamen, H. no date. Fourteen Hundred Years of Bavarian History in Imago Bavariae. Herder. Munich.

Thränhardt, D. (1973) Wahlen und politische Strukturen in Bayern 1848-1953. Droste Verlag. Düsseldorf.

Ücker, B. (1976) Das dritte Königreich Bayern. Süddeutscher Verlag. Munich.

Chapter Three

Bayerische Landeszentrale für politische Bildungsarbeit publication (1993) Die bayerische Verfassung (the Bavarian Constitution).

Bocklet, R. (1979) Das Regierungssystem des Freistaates Bayern. Band 2. Vögel. Munich.

Grundsatzprogramm der Christlich-Sozialen Union in Bayern. In Freiheit dem Gemeinwohl verpflichtet. (1993) (party programme) Atwerb Verlag. Munich.

Hamm-Brücher, H. (1981) Vorkämpfer für Demokratie und Gerechtigkeit in Bayern und Bonn. Liberal-Verlag. Bonn.

Lipset, S. and Rokkan, S. (1967) Voter Alignments: Cross National Perspectives. Free Press. New York.

Mintzel, A. (1975) Die CSU. Anatomie einer konservativen Partei. Westdeutscher Verlag. Opladen.

Mintzel, A. (1977) Geschichte der CSU. Ein Überblick. Westdeutscher Verlag. Opladen.

Rill, B. (1986) Bavaria Felix. Ein Land, das Heimat ist und Zukunft hat. Schulz Verlag. Percha.

Roth, R. (1986) Freistaat Bayern. Die politische Wirklichkeit eines Landes der BRD. Bayerische Landeszentrale für politische Bildungsarbeit. Munich.

Säcker, H. Der Bayerischer Verfassungshof in ibid. Roth (1986).

Chapter Four

Apel, H. (1968) Der deutsche Parlamentarismus. Rowohlt. Hamburg.

Hübner, E. (1978) Wahlsysteme. Bayerische Landeszentrale für politische Bildung. Fifth edition. Munich.

James, P. (1988) "The Bavarian Electoral System", article in Electoral Studies, Volume 7, No. 1, April 1988.

Kuhn, H. et al. Special supplement on the state of Bavaria in das Parlament, no. 11 on 19 March 1983.

Lange, E. (1975) Wahlrecht und Innenpolitik. Entstehungsgeschichte und Analyse der Wahlgesetzgebung und Wahlrechtdiskussion im westlichen Nachkriegsdeutschland. 1945-56. Anton Hain Verlag. Meisenheim am Glan.

Laufer, H. (1981) Das Föderative System der BRD. Bayerische Landeszentrale für politische Bildungsarbeit. Munich.

Siegel, W. (1978) Bayerns Staatswerdung und Verfassungsentstehung 1945-46. Bayerische Verlagsanstalt. Bamberg.

Chapter Five

Backes, U. and Moreau, P. (1993) Die extreme Rechte in Deutschland. Akademischer Verlag. Munich.

Bayerisches Landesamt für Statistik und Datenverarbeitung publication (1986) Die Landtagswahl 1986 von A bis Z.

Dorondo, D. (1992) Bavaria and German Federalism. Reich to Republic 1918-33, 1945-49. St Martin's Press. New York.

Jaschke, H-G. (1993) Die Republikaner. Profile einer Rechtsauflenpartei. Dietz Verlag. Second edition. Bonn.

Kühnl, R. (1967) Die NPD. Struktur, Programm und Ideologie einer neo-faschistischen Partei. Voltaire Verlag. Berlin.

Mintzel, A. (1989) Das dreieinige Machtkartel in Bayern von Staat, Staatspartei und katholischer Kirche, a chapter in Heinrichs, H-J (ed.) F.F. Strauß. Der Charakter und die Masken. Der Progressive und der Konservative. Der Weltmann und der Hinterwäldler. Athenäum. Frankfurt.

Smith, G. Dimensions of Change in the German Party System in Padgett, S. (ed.) (1993) Parties and Party Systems in the New Germany. Dartmouth. Aldershot.

Unger, I. (1979) Die Bayernpartei. Geschichte und Struktur 1945-57. dva. Stuttgart.

Chapter Six

Dehler, K. Interview with the author in Nuremberg on 9 March 1983. Dr Klaus Dehler, now a surgeon, was the youngest member of the Bavarian FDP parliamentary party in 1954 (he was 27), at the time of the Bavarian Coalition of Four. He was Chairman of the Bavarian Liberals from 1964 until 1967, when he retired from politics.

Gow, D. "Edmund Thatcher is no man of straw", article in The Guardian newspaper on 15 January 1994.

Hamm-Brücher, H. Two extensive interviews with the author in the German Foreign Office (*Auswärtiges Amt*) in Bonn in May and July 1982. At the time Dr Hamm-Brücher was minister of state under Genscher in the

SPD/FDP government, led by Helmut Schmidt. In 1954 Dr Brücher, as she was then, was an education spokeswoman for the Bavarian FDP and a driving force behind establishing the Coalition of Four.

James, P. (1985) Liberalism and West German Coalition Politics: the case of the Bavarian Coalition of Four. A doctoral thesis submitted to the University of Loughborough.

Smith, G. (1979) Democracy in Western Europe. Heinemann. London.

Chapter Seven

Ardagh, J. (1991) Germany and the Germans. Penguin. London.

Bavarian State Chancellery publication in English (1991) Information About Bavaria. Munich.

Bayerische Landeszentrale für politische Bildungsarbeit brochure (1992) Freistaat Bayern. Eine kleine politische Landeskunde. Munich.

Mintzel, A. (1977) Die Geschichte der CSU. Ein Überblick. Westdeutscher Verlag. Opladen.

Mintzel, A. (1993) "Die CSU als Forschunsobjekt" in Niedermayer, O. and Stöss, R. (ed.) Stand und Perspektiven der Parteienforschung in Deutschland. Westdeutscher Verlag. Opladen.

Roth, R. (1992) Freistaat Bayern. Politische Landeskunde. Bayerische Landeszentrale für politische Bildungsarbeit. Munich, especially chapter five on the economy, nature and protection of the environment in Bavaria.

Statistisches Landesamt Bayern (1980) Bayerns Wirtschaft gestern und heute – ein Rückblick auf die wirtschafliche Entwicklung. Munich.

Chapter Eight

Bayern Akademie Report for the Hanns-Seidel-Stiftung eV Munich (1991). A detailed report of a study of electoral behaviour at the 1990 Bavarian state election, prepared by the Dr Schumann Research Institute in Bruckmühl.

Clemens, C. (1990) "The CSU and West German foreign policy" in Politics and Society in Germany, Austria and Switzerland, Vol. 2, no. 1/2, Spring 1990.

EMNID polling predictions, reported on German television (n-tv) in January 1994.

Falter, J. and Schumann, S. (1988) "Affinity Towards Right-wing Extremism in Western Europe" in West European Politics, 11 (2) April.

Forschungsgruppe Wahlen Mannheim (1990) A survey of over 20,000 Germans from the old federal states, carried out in December of that year.

Paterson, W. and Southern, D. (1991) Governing Germany. Blackwell. Oxford.

Saalfeld, T. (1993) "The Politics of National-Populism: Ideology and Policies of the German Republikaner Party" in German Politics Vol. 2 no. 2 1993.

Smith, G., Paterson, W., Merkl, P. (ed.) (1989) Developments in West German Politics. Macmillan. London.

Chapter Nine

Burkett, T. and Padgett, S. (1986) Political Parties and Elections in West Germany. Hurst. London.

Engelmann, B. (1980) Das neue Schwarzbuch. Franz Josef Strauß. Kiepenheuer und Witsch. Cologne.

Hopfenmüller, F. and Brügmann, C. (1989) Franz Josef Strauß. Eine kurze Biographie. Stiehl-Druck. Hanns-Seidel-Stiftung. Munich.

More, G. (1994) "Undercover Surveillance of the Republikaner Party: Protecting a Militant Democracy of Discrediting a Political Rival?" in German Politics Vol. 3 no. 2 1994.

Schöll, W. (1988) Franz Josef Strauß. Der Mensch und der Staatsmann. Schulz Verlag. Percha.

Strauß, F. (1989) Die Erinnerungen. Siedler. Munich.

Stern magazine (1988) Special edition (Extra-Ausgabe) "Das war Franz Josef Strauß", devoted to FJS on 7 October 1988.

Chapter Ten

Bayerische Rundschau (1994) Several news reports on Bavarian television of the 1994 European, Bavarian and Federal elections, including interviews with Edmund Stiober and Theo Waigel are referred to.

Focus (1994) A special report by Basisresearch at the end of March 1994 on the strength of CSU support in both Munich and Bonn in the German weekly news magazine Focus (15/94).

Harenberg (1994) Lexikon der Gegenwart. Fakten, Trends, Hintergr nde. Aktuell 94. Lexikon Verlag. Dortmund.

James, P. (1991) "Germany United: the 1990 all-German elections in West European Politics Vol. 14 no. 3 1991.

Padgett, S. (ed.) (1993) Parties and Party Systems in the New Germany. Dartmouth. Aldershot.

Spiegel (1995) article "Blutige Nase" on Monika Hohlmeier and her brother Max Strauß in der Spiegel 13/95.